ACKNOWLEDGMENTS

No book like this one could have been written without the help and influence of very many individuals. They are too many to be named individually, although most are identified in the text. Thank you all.

My American Museum of Natural History colleagues Niles Eldredge, Eric Delson, and Richard Milner have been kind enough to read the manuscript and to offer valuable suggestions. None of them—least of all Eric—will have agreed with all that is said here, but each deserves my gratitude.

Paleoanthropology is above everything a visual science, and good illustration is critical. I have been fortunate indeed to work in the preparation of this volume with Don McGranaghan and Diana Salles. The work of each is identified by initial at the end of the individual figure captions, and my deepest appreciation goes to both, as it does to Jaymie Brauer, who cheerfully chased down the most obscure of references and (less cheerfully) prepared the index.

This volume would never have been begun without the vision and prompting of Bill Curtis, now of Wiley-Liss Publishing. And it would most certainly never have been finished without the persistence, patience, and encouragement of Kirk Jensen, of Oxford University Press. To both I am most grateful, as I am to Carole Schwager for her careful copyediting and to Dolores Oetting for seeing the book through production.

Finally, my appreciation goes once again to the American Museum of Natural History, both for affording me the opportunity to write, and for the incomparable ambience in which I have been able to do it.

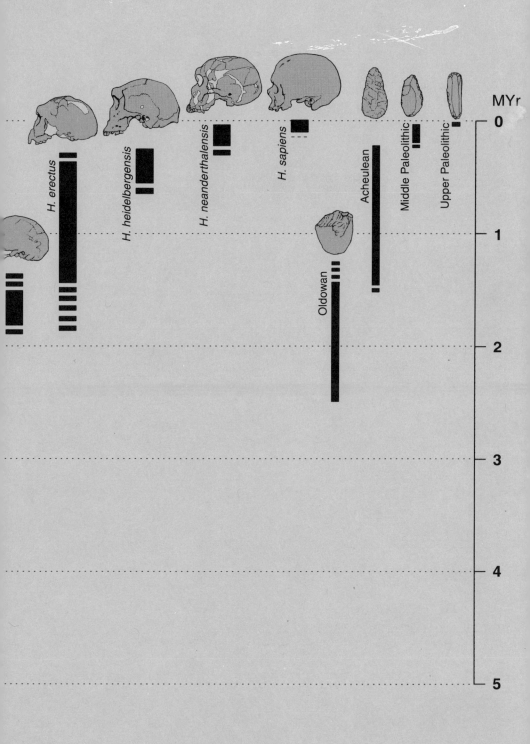

H. erectus

H. heidelbergensis

H. neanderthalensis

H. sapiens

Acheulean

Middle Paleolithic

Upper Paleolithic

Oldowan

MYr

0

1

2

3

4

5

The Fossil Trail

The Fossil Trail

How We Know What We Think
We Know about Human
Evolution

Ian Tattersall

American Museum of
Natural History

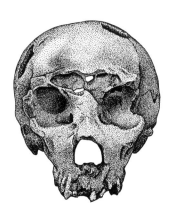

New York   Oxford
OXFORD UNIVERSITY PRESS
1995

Oxford University Press

Oxford   New York
Athens   Auckland   Bangkok   Bombay
Calcutta   Cape Town   Dar es Salaam   Delhi
Florence   Hong Kong   Istanbul   Karachi
Kuala Lumpur   Madras   Madrid   Melbourne
Mexico City   Nairobi   Paris   Singapore
Taipei   Tokyo   Toronto

and associated companies in
Berlin   Ibadan

Library of Congress Cataloging-in-Publication Data
Tattersall, Ian.
The fossil trail : how we know what we think we know about human
evolution / Ian Tattersall.
p. cm.
Includes bibliographical references and index.
ISBN 0-19-506101-2
1. Human evolution. 2. Fossil man. 3. Anthropology, Prehistoric.
I. Title.
GN281.T357   1995
573.2—dc20   94-31633

9 8 7 6 5 4 3 2 1

Printed in the United States of America
on acid-free paper

PREFACE

Sir Isaac Newton once said that if he had seen farther, it was because he had stood on the shoulders of giants. In these uncharacteristically gracious words he acknowledged a debt to the past that is universal among scientists—as well as (presumably unintentionally) a burden that is equally universal. For although every scientist starts from a base established by his (or her) predecessors, what you see from your lofty elevation depends on how tall your giant is, and in what direction he happens to be facing. That's what this book is about, for how you read your evidence is at least partly conditioned by what you are expecting to find; and in the science of paleoanthropology preconception may well have played an even larger role than in most other sciences. Of course, the study of human evolution has come a long way since its early days, in terms both of the basic fossil evidence and of how it is analyzed. But we are still largely in thrall to received wisdom, and this brings us back to the central theme of this book. How—and why—have we come to know what we think we know about human evolution: about the complex history of our own biological past?

Most popular books about human evolution in recent years have been based on the experience of individual paleoanthropologists in the field, and thus have at least implicitly projected the notion that reconstructing the past is essentially a matter of discovery: find enough fossils, and all will be revealed. This in turn reflects the idea that paleontology is somehow like a giant jigsaw puzzle, and that once we have all the pieces they will fit together to disclose the full picture; or at the very least, that when we have enough pieces we will be able to discern the broad outlines of the design.

Hence the traditional paleontologists' lament: the inadequacy, almost invariably described as "woeful," of the fossil record. Well, it's true that we will never have a "complete" fossil record. In fact, we will never have a human fossil record that preserves even one thousandth of one percent of all the individuals that have ever lived. But even now we have a reasonably good sampling of fossil species—even fossil human species—that should allow us, by appropriate analysis, to gain a provisional idea of the major events that led to the emergence of our own kind on Earth. I use the term "provisional" in a positive sense, because all scientific knowledge is provisional; indeed, how can we expect to make progress in any area of science if what we believe now is not somehow inaccurate or at least incomplete? A scientific idea is one that can be tested in the light of new observations, whether these new observations are experimental or are based on new discoveries or on new analyses of old discoveries. Popular misconceptions to the contrary, scientific ideas are not immutable declarations of truth, nor are they intended to be.

But the starting point for any new set of hypotheses is the set of hypotheses that preceded it; and what we believe today can never be fully independent of what we believed yesterday. Moreover, in anything as close to our own ego as the story of our own origins, what we think we know cannot be independent of what we believe about ourselves. Clearly, it is too much to ask that scientific opinions in this emotive realm should be entirely independent of prevailing social thought and attitudes. So in trying to comprehend how we know what we think we know today about our evolution, it's important to look back at the past of paleoanthropology and to understand by what circuitous routes we have arrived at that knowledge. What we have believed in the past, the evidence we have now, and how we look at that evidence all interact in a complex way. And that is why this book follows a historical path.

New York   I. T.

CONTENTS

ABBREVIATIONS

AMNH	American Museum of Natural History
DK	Douglas Korongo (a locality at Olduvai Gorge)
FLK	Frida Leakey Korongo (a locality at Olduvai Gorge)
FLKNN	FLK North—North (a locality at Olduvai Gorge)
KBS	Kay Behrensmeyer Site (at East Turkana)
KNM	Kenya National Museums
KNM-ER	Kenya National Museums—East Rudolf
KNM-WT	Kenya National Museums—West Turkana
LH	Laetoli Hominid
MLD	Makapansgat Limeworks Deposit
MNK	Mary Nicol Korongo (a locality at Olduvai Gorge)
NME	National Museum of Ethiopia
NME—AL	National Museum of Ethiopia—Afar Locality
NMT	National Museum of Ethiopia
NMT-WN	National Museum of Ethiopia—West Natron
OH	Olduvai Hominid
SK	Swartkrans
Sts	Sterkfontein site (TM designation)
Stw	Sterkfontein site (UW designation)
TM	Transvaal Museum
UW	University of the Witwatersrand

The Fossil Trail

1

Before Darwin

Interest in our own origins dates back to a time well before anyone realized that we had a fossil record, or even an evolutionary past. Indeed, every human society has its own origin myths, reflecting the need for self-explanation that seems to be so deeply ingrained an aspect of the human psyche. Nonetheless, Western scientists first studied human fossils for much the same reason that mountaineers have always scaled peaks: simply because they were there. Of course, by the mid-nineteenth century, when the human fossil record began to yield up its secrets, there was already intense interest in the remarkable diversity of the living world and in humankind's place within it. This was matched by an increasing awareness that the world had to be a much older place than Scripture (at least as translated into seventeenth-century English) appeared to suggest. Actual fossil humans, though, were co-opted rather late into evolutionary and geological schemes; and many of the early explanations for the differences between ourselves and, say, the original Neanderthaler (discovered in 1856) had a somewhat ad hoc flavor to them. Certainly, reading these imaginative speculations today one hardly gets the impression that they were informed by any body of theory or expectation, coherent or otherwise. An unkind observer might remark that over the past century and more the practice of paleoanthropology—the study of the human fossil record—has continued to plough a furrow rather distinct from that of mainstream evolutionary biology; but perhaps a certain insularity is inevitable in a study that touches ourselves as closely as this one does.

By the eighteenth century zoologists were becoming increasingly preoccupied by a fact already apparent to the ancients: there exists in nature an order such that humans resemble some nonhuman creatures much more than others. A relatively typical viewpoint of the period was that of Karl von Linné (better known as Carolus Linnaeus), the Swedish savant who created the system of naming and classifying living things that we still use today. Linnaeus's system, much elaborated in the past couple of centuries but still intact in its essentials, consists of a hierarchy of increasingly inclusive categories. Thus Linnaeus grouped species into genera, genera into orders, and orders into kingdoms, the largest subdivision of living things.

In the order Primates, along with *Vespertilio*, the genus of the bats, *Lemur*, the genus of the Malagasy lemurs and various allied forms, and *Simia*, the genus of the monkeys, Linnaeus included the genus *Homo*. In addition to human beings, placed in the species *Homo sapiens*, this last genus contained *Homo troglodytes*, a species that embraced both the great apes known to Linnaeus: the chimpanzee and the orangutan. Hardly a ripple greeted the publication of this part of Linnaeus's classification; after all, it was undeniable that of all living animals the apes most closely resembled humans, a fact that had been elegantly demonstrated years before by the great English anatomist Edward Tyson. In comparing the anatomy of a chimpanzee with that of a human and a monkey, Tyson had shown in 1698 that the ape and the human resembled each other in forty-seven respects, while the ape resembled the monkey in only thirty-four. No problem: as long as the idea of the fixity of species was not attacked, as long as the pattern of resemblances between species in nature was accepted as simply reflecting the will of the Creator, the simple recognition of similarities evident to the most inexpert eye posed no threat at all to the prevailing beliefs of the day. Human beings were still distinct, separately created from the rest of nature—created, indeed, to give nature meaning. A world without people, most Europeans believed, would be a world without purpose.

Not that anyone thought the world had lacked purpose for long, even under the biblical account of its creation. A literal interpretation of the word "day" in the translations of Genesis implied that humans had entered the world only fractionally after its creation, for a week is but an eyeblink in the approximately six thousand years which theologians broadly agreed had elapsed since God had placed Adam on Earth. Martin Luther himself had subscribed to this figure, and numerous theologians had come up with similar ages for humankind. The date of approximately 4000 B.C. was arrived at by totting up the genealogies recounted in the Old Testament and fleshing them out with astronomical calculations of various kinds: a methodology that in some hands allowed calculations of astonishing precision. Thus John Lightfoot, vice-chancellor of the University of Cambridge, concluded in the mid-seventeenth century that the Creation had occurred at 9 A.M. on October 23, 4004 B.C.—conveniently at the beginning of the academic year!

But an equally beguiling aspect of theology is that, like constitutions, sacred documents usually leave ample room for interpretation according to taste. By the waning years of the eighteenth century some naturalists were beginning to make use of loopholes in the Book of Genesis to envisage a rather longer time span than six thousand years since the creation of the world, even though nobody as yet dared suggest that this figure was not roughly right for the origin of mankind. Events surrounding the French Revolution spurred tendencies of this kind in France, particularly, where a new, theistic view became fashionable. This held that the Creator had started the ball rolling but had since more or less stayed on the sidelines, observing the unfolding of events presumably with some degree of interest but without much active involvement—at least until humans came upon the scene. The great French naturalist Comte Georges de Buffon used this emerging interpretation to the greatest advantage. Structuring his account of the history of Earth along the lines of Genesis, he described seven phases of life, with humans appearing only in the sixth, and human modification of the natural world occurring only in the last. Thus Buffon envisaged a dynamic, constantly changing earth that had been in existence for tens of thousands of years. That relatively long period was essentially one of preparation for mankind, which appeared only when the world had become "worthy of its rule."

Through the efforts of Buffon and others, by the end of the eighteenth century the idea that Earth itself had had a long history was widely accepted, even though this meant that the account in Genesis had to be read as allegory. The biblical genealogies still ruled the day, however, and the six thousand-year date for the creation of humankind continued to be accepted, by Buffon as by most others, as a good approximation.

If Earth indeed had a long history, then geological observations could not be kept apart for long from discussions of that history. Before the turn of the nineteenth century another outstanding French scientist, Georges Cuvier, drew attention to the fact that the geological record is not a continuous one; indeed, it is marked by numerous dramatic breaks. He also noted that in any one place sediments laid down in seas might be succeeded by terrestrial deposits, only to be capped in their turn by more marine strata. Clearly, the sea had invaded the land on many occasions. Moreover, he saw that the lower down in a sequence of strata he looked, the less like the modern fauna the fossils contained in the rocks became. He demonstrated the fact of extinction (which was resisted by many as incompatible with the perfection of Divine Creation) by his studies of huge fossil mammals that could not possibly have been overlooked by describers of the contemporary fauna if they still existed on Earth. Such fossils of ancient animals were abundant in geological deposits near Paris. These deposits consisted of superficial gravels that geologists of the time associated only with vigorous water action—though we now know that they were due to glaciation. Putting all these facts together, Cuvier came up with the idea that the history of the world was marked by a series of "revolutions" or "catastrophes," in which entire faunas were wiped out as land was flooded and seafloor exposed. Bowing to received wisdom, Cuvier suggested that the most recent of the extinct faunas he had described might date back to about six thousand years ago; however, it was not in France but in England that the closest parallels were drawn between this most recent catastrophe and the Noachian flood.

Cuvier himself was always scrupulous in limiting his arguments to the evidence at hand and in avoiding the introduction of theological considerations into them. His famous statement that "fossil man does not exist!," for example, referred simply to the fact that there were no human fossils known to him; it was later interpreters who added the implication that such *could* not exist. But even before Cuvier's major work on catastrophism appeared in 1812, the English physician James Parkinson, best known for describing the disease that bears his name, had seized upon his catastrophist ideas as scientific confirmation of the biblical account of creation. The idea that the creation of humankind represented the culmination of the divine plan, as revealed in Genesis, was further confirmed by the failure (up to that time) of paleontologists to find human bones in association with the extinct faunas. Further, the existence of the gravels containing extinct animals became evidence for the Deluge. But although this theme was taken up by a number of influential scholars such as the eccentric but brilliant geologist and divine William Buckland, it did not long go uncontested. The science of geology was developing apace, and Buckland's arguments were rapidly demolished by the likes of the geologist Charles Lyell, who was once described as having a mission to "free science from Moses."

Lyell it was who, in his *Principles of Geology* (1830), established the basic chronology of what we now know as the Tertiary Period, roughly the last sixty-five million years. Lyell divided this stretch of time (then, of course, unquantified in years) into, successively, the Eocene, Miocene, and Pliocene epochs. In a later edition of his book Lyell appended the Pleistocene epoch to the end of the series, and other authors subsequently added the Oligocene between the Eocene and Miocene, and the Paleocene at the beginning of the series. Not long after Lyell introduced the concept of the Pleistocene, the paleontologist Hugh Falconer noted that it coincided with the Ice Ages that had been identified first in Europe and then in America by the Swiss-born geologist Louis Agassiz. The "diluvial" gravels were eventually identified as the results of one or another of the episodes of climatic cooling and consequent glaciation—expansion of the polar ice caps—that occurred

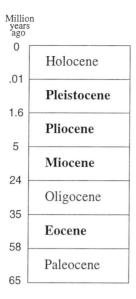

Million years ago	
0	Holocene
.01	**Pleistocene**
1.6	**Pliocene**
5	**Miocene**
24	Oligocene
35	**Eocene**
58	Paleocene
65	

Sequence of geological epochs over the past sixty-five million years, as dated by modern methods. The Paleocene through Pliocene together make up the Tertiary Period; the Pleistocene and the Holocene compose the Quaternary Period. Epochs named by Lyell are shown in boldface type.   DS.

during the Pleistocene. Much later, the concept of the "Holocene," or "Recent," was introduced to cover the last ten thousand years or so since the climate warmed up at the end of the last glacial episode. The unpredictable results of human interference with such patterns aside, however, there is in fact no reason to think that we are out of the glacial cycle that characterized the greater part of the Pleistocene.

A major upshot of the routing of Buckland by the mainstream geologists was the effective expunging of the biblical chronology from scientific discussion after the 1830s. If the diluvial gravels could not be associated with specific biblical events, then the relevance of the biblical timetable to the geological record vanished. Although this chronology was not forgotten by a generally devout populace, the way was open for geologists to contemplate a truly enormous antiquity for a world. As James Hutton, one of the fathers of geology, had already put it back in 1795, there was "no vestige of a beginning, no prospect of an end."

Geologists were not, of course, centrally interested in establishing the antiquity of humankind. Indeed, in the 1830s they had no evidence—or, at least, only negative evidence—to bring to bear on the matter. The antiquarians, in contrast, had materials in hand that were highly germane to the issue. Imperishable as they are, stone tools form the most abundant category of evidence about the activities of early humans; but it was necessary first to recognize them as human artifacts. Flaked flint tools, known in Europe from time immemorial, were acknowledged to be curiosities that required explanation. Such explanations were as varied as they were imaginative: petrified thunderbolts, fairy arrows, exhalations of the clouds. In Europe, where (despite the use of gunflints) it was unthinkable by then to make cutting or scraping tools from stone when so many other superior materials were available, it did not occur to anybody that these strange objects might have a prosaic explanation—until, that is, the New World was discovered, for there people still did make and use stone tools. At that point the ingredients, if not the intellectual environment, were there to make the correct diagnosis. The first person to publish the idea that the "thunderbolts" were the artifacts of ancient "pre-Adamite" humans was Isaac de la Peyrère of Bordeaux, in 1655. Alas, poor Isaac suffered for his temerity, finding himself seized by the Inquisition and his book publicly burned on the streets of Paris.

In England a more tolerant if not more receptive social climate reigned. Shortly after de la Peyrère's book appeared Sir William Dugdale, author of a survey of the antiquities of Warwickshire, illustrated a stone axe which he claimed was made by ancient Britons who lacked the use of metals. Dugdale's proposition was generally ignored, but several similar suggestions appeared in the years following, and during the eighteenth century, a period somewhat less doctrinaire than the seventeenth, several French scholars, including the renowned botanist A. L. de Jussieu, took up the theme that stone tools were the work of people who knew nothing of metals. Sadly, however, the significance of the first discovery of extinct animals in association with human bones and stone tools, in the Gaylenreuth Cave in Germany, was denied by its discoverer. In 1774 Johann Friedrich Esper wrote that he "dared not presume" that the human remains at the site were contemporaneous with the animal fossils among which they were found. It was thus not until the end of the eighteenth century that John Frere, an English country squire, more or less correctly diagnosed a group of Stone Age handaxes found twelve feet down, and once more in association with the bones of extinct animals, in a gravel pit at Hoxne, in Suffolk. These, he wrote in 1800, were weapons made by people who knew nothing of metalworking; and their geological context "may tempt us to refer them to a very remote period indeed, even beyond that of the present world." By which he meant, of course, the six thousand year world of biblical chronology.

Prescient as they were, Frere's remarks went as unheeded as their predecessors. So did the first discovery of stone tools and the bones of extinct animals in association with an extinct kind of human. This occurred about thirty years later at the cave of Engis, in a valley wall of the River Meuse, near Liège in Belgium. The excavator of the site, Philippe-Charles Schmerling, recognized the artifacts as the work of ancient peoples; but when he found himself unable to pay his printer, most of the copies of his great work on the *Bones of the Caves of the Province of Liège* were sold for scrap paper, severely limiting its availability to scholars. Moreover, of the two human skulls that Schmerling found, one, though of quite early date, had belonged to an individual of modern form; and the older one, though the remains of an extinct kind of human, looked deceptively modern because it had belonged to a child. Finally, Schmerling's premature death and the transfer of his Engis collections to the custody of the University of Liège, which stored them in the back of a barn, almost guaranteed obscurity for his discoveries. It is indeed remarkable that they survived to attract the attention they belatedly received. Although the great age of the human remains was never seriously contested—and was eventually confirmed by no less an authority than Sir Charles Lyell, who accepted their association with extinct species such as mammoths and woolly rhinoceroses—it was nonetheless many years before their true significance was recognized. A similar fate awaited an adult and much more complete archaic skull found in Gibraltar during work on military fortifications in 1848 or earlier; this also sat neglected on a shelf for many years before its importance became understood.

It is thus Jacques Boucher de Perthes, a customs official from Abbeville in northern France, who is usually credited with putting stone age prehistory on a firm footing. In the late 1830s, after unsuccessfully striving for acclaim as a poet and playwright, Boucher de Perthes turned his interest to the stone tools that were being found in various places in the Somme river valley, some of them in association with extinct faunas. At first he believed that these objects were diluvian, but he rapidly abandoned Noachian explanations and came to refer to them in his voluminous works as antediluvian—from before the Flood. At that time most geologists in France were still catastrophists, and Boucher de Perthes ran into a general climate of incredulity which was fueled by his tendency to be a little undiscriminating about what was and was not a tool or a fossil. He held his ground tenaciously, however, and almost from the beginning found some support among French scientists. Eventually many were won over to his view that these objects were both tools and, as testified by the associated fossils, truly ancient. Similar work had been under way in England, and the turning point for Boucher de Perthes came in 1859, when two British scholars, Joseph Prestwich and John Evans, visited the Somme valley and authenticated his finds. On his return Evans reported as much to the Royal Society,

Chert Acheulean handaxe (AMNH 6.9) from the Somme River valley, France, similar to those described by Jacques Boucher de Perthes. Scale is 1 cm.   DS.

placing a quasi-official imprimatur upon the ancientness of mankind. Fittingly, that year was also the year in which Charles Darwin published *On the Origin of Species*.

As the antiquarians were gradually demonstrating the long history of humanity, the science of paleontology—the study of ancient life through the fossil record—was also becoming established, not least through the efforts of Cuvier. But proving the existence of extinct species was one thing; admitting the transformation of one species into another was something else entirely. Cuvier, like virtually all others of his day, perceived species as static, unchanging entities, and nothing that he saw in the fossil record appeared to him to violate this conventional point of view. Faunas certainly succeeded each other in that record, but this was due either to the immigration, from elsewhere in the world, of animals that had escaped destruction in the previous catastrophic episode, or to re-creations following each catastrophe. In the latter case Cuvier declined to be too specific, for, excellent scientist that he was, he did not like to mix religion and science. He was not, however, prepared to contemplate the idea that one fauna—or species—could change into another.

Cuvier's beliefs were enormously influential. Elaborated by a legion of followers, they continued to dominate French geology and paleontology for decades after his death in 1832. Nonetheless, his approximate contemporary Jean de Monet, Chevalier de Lamarck, came to believe otherwise. Cleaving for most of his career to the standard view of the fixity of species, Lamarck underwent a sort of Damascene conversion between 1799 and 1800 that radically changed his viewpoint. This change apparently germinated in his study of a collection of mollusks, in which he found that many fossil species appeared to have counterparts in the living fauna—hardly what one would expect under Cuvier's catastrophism. Moreover, he found that in many cases he could arrange his mollusks into a modifying series that seemed gradually to proceed from older fossil species to younger extinct species to living ones. Deducing that such series represented connected lineages composed of ancestors and descendants, Lamarck in turn concluded that species could indeed change, slowly, from one into another. More generally, he decided this meant that the enormous diversity of species in the living world was due to the gradual divergence of lineages over vast spans of time. He thought that such lineages kept changing to keep pace with changing environments. And Lamarck had no problem with time: unfettered by biblical considerations, he was prepared to contemplate a limitless history of Earth. Moreover, he was even prepared to speculate that humans had arisen through this process from an apelike animal which adopted upright locomotion. These conclusions were comprehensively presented to the world in 1809, when Lamarck published his *Philosophie Zoologique*.

Having glimpsed the essential outline of the history of life, and having replaced a static concept of the living world by a dynamic one, the question Lamarck faced was this: if lineages change, how do they do it? Unfortunately, his choice of answer was to destroy his reputation among future historians of science. First, he proposed a sort of inbuilt tendency for organisms to become more complex, although he recognized that the actual history of life as seen in the fossil record was untidier than this would imply. An additional ingredient was needed. Since species had to be in harmony with the environment, yet environments fluctuated, species had to have a way of changing as environments changed. Lamarck proposed, therefore, that new behaviors elicited by changing environments caused changes in the organisms themselves. The new characteristics acquired as a result of these changed behaviors would then be passed along to offspring, and gradual physical modification of the species would ensue. The starting point of this process was achieved by a sort of spontaneous generation—never precisely defined—of the ancestral species.

In contrast to his ground-breaking observations of change within lineages and the mutability of species, the twin notions of the modification of organs by use or disuse and of the inheritance of acquired characteristics, were not new with Lamarck. But it is with them, and particularly with the latter, that Lamarck's name has in hindsight become associated; and by extension the entire body of his work has appeared discredited. In his own day, Lamarck's ideas were widely and savagely attacked by scientists as diverse and influential as Cuvier in France and Lyell in England; with the rediscovery of the principles of genetics at the beginning of this century, his borrowed wrong ideas on the mechanism of change caused him to be reviled all over again. The result is that the baby vanished with the bathwater; but had it been Lamarck's rather than Cuvier's voice that had been more loudly heard at the beginning of the nineteenth century, the course of evolutionary thought might have been very different.

As it was, Lamarck's insights were ignored where they were not opposed; even among the uniformitarian geologists such as Lyell—who were, after all, busy proving that the world had in the past been shaped by the same forces that we see operating today—it continued to be believed that Earth and its contents were ultimately the handiwork of a deft Creator. It caused more than a stir, therefore, when in 1844 the London encyclopedist Robert Chambers published—anonymously—a work called *Vestiges of the Natural History of Creation*. Little wonder that Chambers preferred to remain anonymous: although he accepted the divine Creator standing behind the history of life on Earth, he proposed that life evolved through time, as the result of gradual changes that had nothing to do with catastrophes of any kind, and according to a "principle of progressive development". Many of the points he made anticipated those that Charles Darwin was to make fifteen years later; but as a nonscientist Chambers also made a variety of blunders, and in assembling his evidence he failed to discriminate consistently between folklore and scientific fact. What's more, Chambers failed to come up with a convincing "engine" for evolution. Nonetheless, his book was a remarkable piece of work, even though it provoked more outrage than re-examination of comfortable British prejudices. And it kept alive in the minds of biologists the threat of Lamarckism, of the possibility that life had indeed changed through time. Those who studied living things in the mid-nineteenth century could not totally ignore the possibility that evolution underlay the diversity of the living world, even if in the best of faith they denied it.

In Germany, meanwhile, the first half of the nineteenth century witnessed the growth of *Naturphilosophie*, a romantic and occasionally rather wild-eyed movement whose adherents were reluctant to accept purely mechanistic explanations for natural phenomena, though not all of them took refuge in theological explanations of cause. Many espoused the notion of some inner impetus toward "development" in organisms, either through

—

Side view of the original Neanderthal skullcap found near Düsseldorf, Germany, in 1856. Type specimen of Homo neanderthalensis. *Scale is 1 cm.*   DS.

the expression of some preexisting potentiality or through some form of transmutation. By 1851 Chambers's book had been translated into German and had influenced such important thinkers as Schopenhauer. So German science during the 1850s was in a sense primed to be receptive to darwinian ideas when they were finally published. Indeed Hermann Schaaffhausen, the principal describer of the Neanderthal fossils, wrote in 1853, before these fossils had been found and before Darwin's book had appeared, that "the immutability of species . . . is not proved." The title of Schaaffhausen's article, "On the Constancy and Transformation of Species," reflected an active debate of the time in German scientific circles.

It was this intellectual milieu that greeted the first human fossils recognized as sufficiently remarkable to justify a special effort at explanation. In 1856, workers began clearing out a small cave in the steep side of the Neander Valley (Neander Thal), through which the river Düssel reaches the Rhine. Inside, they uncovered a skeleton buried below some five feet of mud. A few bones of what may when found have been an intact skeleton survived their disinterment and came to the attention of the local savant Johann Fuhlrott. Correctly recognizing them as both ancient and human, Fuhlrott passed them along to Schaaffhausen, a professor of anatomy in Bonn, for fuller description. The surviving bones consisted of a skullcap, both femora (thighbones), elements of the upper and lower left arm, part of the pelvis, and a few other bits and pieces.

In many ways it is hard not to be impressed by Schaaffhausen's report, published in 1858 and translated into English by the anatomist George Busk in 1861. Schaaffhausen gave a minutely detailed anatomical description, in the course of which he remarked on the thickness of the bones and on the large size of the scars left by the muscles that had attached to them. These suggested to him that the Neanderthal individual had been extremely strong and muscular, possibly as a result of a very strenuous lifestyle. Most of all, however, Schaafhausen's attention was drawn to the unusual shape of the skullcap, and particularly to the presence of the pronounced ridges above the eyes, the large frontal sinuses, and the low, narrow forehead, which together gave "the skull somewhat the aspect of that of the large apes." These features, which he concluded were the results neither of artificial deformation nor of pathological deformity, placed the specimen beyond anything with which he was directly familiar. But what did they mean?

"Sufficient grounds exist," Schaaffhausen wrote (as translated by Busk), "for the assumption that man coexisted with the animals found in the *diluvium*; and many a barbarous race may, before all historical time, have disappeared, together with animals of the ancient world, whilst the races whose organization is improved have continued the genus." It thus made eminent sense to search for something comparable to his specimen among the numerous existing reports of human skulls found in archaeologically ancient contexts. This he did at length, even mentioning that the skull from Engis was reported to have a similarly narrow frontal bone—though he was referring to the modern-looking adult skull that Schmerling found there; and, in any event, the truly archaic child's skull from Engis was too young to show the brow ridges that it would have acquired had the individual survived to adulthood. Although he was thus unable to identify anything in the literature that quite matched his Neanderthaler, Schaaffhausen found as a result of his survey "that a marked prominence of the supraorbital region, traces of which can be perceived even at the present time, occurs most frequently in the crania of barbarous, and especially of northern races, to some of which a high antiquity must be assigned, [so] it may fairly be supposed that a conformation of this kind represents the faint vestiges of a primitive type, which is manifested in the most remarkable manner in the Neanderthal cranium." Schaaffhausen then went on to document the savagery of various ancient European tribes (mostly as seen through the eyes of classical or early Christian chroniclers) and to conclude that, along with various others, his specimen "may probably be assigned to a barbarous, aboriginal people, which inhabited the north of Europe before the *Germani*." Nonetheless, it is clear that Schaaffhausen was not entirely happy with this conclusion, for "the human bones from the Neanderthal exceed all the rest in those peculiarities of conformation which lead to the conclusion of their belonging to a barbarous and savage race."

Schaffhausen's analysis is a classic example of a careful, insightful and scholarly analysis confined by the limitations of received wisdom. By 1858 the fact that recognizable humans had first appeared in the remote past was beyond doubt; and the progressionist views prevailing in Germany made it easy to envisage that, physically or physiognomically, ancient savage tribes might have differed somewhat from civilized humans; indeed, Schaaffhausen claimed that "even in ancient times the various Germanic stocks . . . in proportion as they led a savage or more civilized mode of life, differed in corporeal constitution, as well as in the formation of the face and head." But as long as the idea of the mutability of species (as opposed to varieties of species) was entertained only as an abstraction, as a philosophical concept rather than an organizing principle of the diversity of life, what Schaaffhausen and everyone else necessarily lacked was the idea of common descent among species (although he could think in analogous terms about "tribes"). That the Neanderthaler, so evidently human but equally evidently unlike any other human known, might represent not a variety but a relative—perhaps even an ancestor—of modern humankind was totally foreign to the prevailing mindset. It was simply too great a leap to make on the basis of a single incomplete and rather odd-looking skull. Given that Schaaffhausen somehow had to fit the Neanderthal remains within a view of nature that we know with the advantage of hindsight was not appropriate, it is remarkable that he was able to achieve as balanced an analysis as he did.

In commenting on Schaaffhausen's review of this unprecedented material, Busk himself stressed the comparison between the Neanderthaler and the great apes. He also drew attention once again to comparisons with the adult Engis skull—ironically, the modern one and not the infant cranium, which, as we know now, is indeed that of another Neanderthal. But a colleague of Schaaffhausen at the University of Bonn, Friedrich Mayer, took issue with him on every point. Writing in 1864, five years after Darwin's publication of *On the Origin of Species*, and at a time when the potential significance of the Neanderthaler was already clear to him, Mayer pointed out that this skull lacked a sagittal crest—the bony ridge atop the braincase to which massive jaw muscles attach in large apes such as male gorillas. We know now that a sagittal crest is merely a passive result of combining big chewing muscles with a small braincase; and the Neanderthal braincase was a large one, containing a brain as big as our own. At the time, however, this was less apparent. "Show me a fossil human skull with a sagittal crest," Mayer demanded, "and I will admit that man descends from an apelike ancestor."

Thus refuting Schaaffhausen's interpretation, Mayer proposed two alternatives. The first was that the Neanderthal remains were those of a modern individual with pathological degeneration of the skeleton caused by childhood rickets. The shape of the femora suggested to him that the Neanderthaler was bow-legged, as many horsemen become, and he finally concluded that these were the remains of a deserter from the Cossack army that had paused in the area of the Neander Valley before crossing the Rhine in January 1814 on its way to invade France. There were also signs of trauma, duly noted by Schaaffhausen, in the surviving elbow joint of the skeleton. The individual thus suffered constant pain; and this, Mayer claimed, had combined with the sequelae of the rickets to cause a permanent frown. The strained facial muscles in turn had affected the shape of the eyebrow region! This reasoning sounds very quaint and amusing today, but at the time it was a theme taken up by many. Over the longer haul, though, it was Mayer's alternative suggestion to explain the peculiarities of the specimen—that the Neanderthal skull had belonged to a modern individual showing pathological changes—that received most support.

What I find particularly curious about this episode, though, is that in his 1858 paper Schaaffhausen acknowledged that Professor Mayer had early on drawn attention to dendritic encrustations on the fossil—and it was precisely on the basis of these that Fuhlrott had eventually concluded that the specimens were indeed fossilized. Mayer may already have been hesitant about Fuhlrott's and Schaaffhausen's interpretation of the bones; but it nonetheless seems likely that Mayer's preferred interpretation—which had a dying Cossack climbing up sixty feet of near-vertical cliff to bury himself, naked, under five feet of mud—may have had at least as much to do with the politics and personalities of the Bonn anatomy faculty as with the attributes of the specimen itself. If so, *plus ça change, plus c'est la même chose.*

2

Darwin and After

As we've seen, Charles Darwin's *On the Origin of Species* did not burst forth into a world that was totally unprepared for the concept of evolutionary change—unwilling to accept it, perhaps; but hardly a world that was unfamiliar with most of the evidence that Darwin used to support his ideas. Certainly by 1859, the year of the great book's publication, there was more than enough evidence around to convince any open-minded observer of the reality of evolution. Indeed, the naturalist Alfred Russel Wallace had already come to a conclusion strikingly similar to Darwin's in a manuscript he sent back to England from the remote island of Ternate in 1858. This was read, together with extracts from Darwin's writings, at the Linnaean Society that year. But remarkable though it may seem in retrospect, at the time few might have disagreed with the Society's president, who opined in his annual summing up of events that "The year has not been marked by any of those striking discoveries which . . . revolutionise . . . the department of science on which they bear." The idea of evolution was in the air in 1859, but it was not yet an idea whose time had irresistably come. What won the day for *On the Origin of Species* was not merely that Darwin was a respected naturalist to whom his peers would listen, but that in his book he documented his ideas in eloquent, exquisite and exhaustive detail. He produced a work which, quite simply, could not be ignored. It is certainly true that from the vantage point of the late twentieth century we may tend to overestimate the influence of Darwin's ideas of natural selection on late nineteenth and early twentieth century evolutionary thought; but there is no doubt that, despite the resistance he initially encountered, it was Darwin who made the concept of evolution respectable, and who set the framework for our modern understanding of how life on earth came to be the way it is.

Darwin has been the subject of numerous careers in the history of science;

and he wrote so voluminously that by judicious quotation one can defend almost any evolutionary viewpoint as somehow "darwinian." Nonetheless, the essentials of Darwin's theory of evolution can be quite simply expressed. His own capsule definition of evolution was "descent with modification": ancestral species give rise to descendants which do not exactly resemble them. In searching for a mechanism that would explain how this might occur, Darwin faced a politicoreligious problem as well as a scientific one: he had not only to come up with a plausible means of change, but he had somehow to destroy the pervasive notion of the fixity of species. Darwin's solution was simple. His studies of barnacles had already suggested to him that species were difficult or even impossible to recognize in the living fauna, with blurred boundaries between close relatives; but he delivered the *coup de grâce* to their fixity by denying them identity in time. Here's how.

In every generation, as naturalists had long been aware, individuals of the same species vary from each other in numerous ways. Some of those individuals are better adapted than others to survive in any given environment. At the same time, in any generation many more individuals are born than will ever survive to grow up and reproduce themselves. It is the better adapted who will reproduce most successfully, and, since traits are inherited, they will pass their superior adaptations on to their offspring. Each generation will thus be slightly different from the last, with more individuals possessing favorable adaptations and fewer having less well adapted anatomies. In this way, over long spans of time, lineages will gradually become modified as they adapt to changing environments (or as they perfect their adaptations to existing environments) by the process Darwin named "natural selection."

This mechanism of change is not, Darwin stressed, in the least purposive; it consists merely of a sort of winnowing action achieved through the vagaries of the environment. Less well adapted individuals contribute fewer offspring (resembling themselves) to the next generation, either through early mortality or through reduced reproductive effectiveness; those more favorably endowed live and produce more offspring (again resembling themselves). Most simply stated, then, natural selection is differential reproductive success. It is central to this idea that offspring inherit their particular traits, advantageous or disadvantageous, from their parents: a fact that had been established since time immemorial by animal breeders, and that was obvious to anyone who had ever noted the fact of familial resemblance. Interestingly, however, Darwin had no idea how this occurred. Indeed, Darwin's theories of inheritance were totally wrong, though he has escaped the pillorying for this deficiency that the unfortunate Lamarck has suffered. What is important, though, is that Darwin was perfectly able to formulate his enduring ideas on evolution *in the absence* of a valid theory of inheritance. This is a theme that we will return to later.

Now, while in one respect Darwin's ideas on evolution by natural selection ran counter to prevailing views of the fixity of species, in another way they were highly congenial to the spirit of the times in which he lived. For the Victorian ethos was one of progress, of improvement. Leave aside that the working classes labored in conditions of appalling privation; to those with the leisure to think about such matters the ubiquity of progress was evident. The explosion of technological change and economic expansion unleashed by the Industrial Revolution reached into all areas of nineteenth-century experience to inculcate a brash and optimistic view of the inevitability of progressive change. And this change was, of course, of an incremental, gradual sort: revolution was unthinkable, and certainly undesirable. In this sense, Darwin's ideas were very much in tune with his times: whether consciously or not, he adapted to biology a more generally held notion of the inevitability of progress.

Nonetheless, once the furor that greeted the publication of his book died down, it was for his gradualistic views on natural selection that Darwin was most strenuously criticized. Indeed, they continue to this day to be the focus of critical attention. In contrast, however, Darwin's central thesis, still the linchpin of modern evolutionary theory, was quite rapidly accepted by his peers—though it was immediately more shocking to the general public. This central thesis is that all forms of life are related by ancestry. Darwin perceived the same pattern in nature that Aristotle and others had noticed in classical times, and that had been rediscovered in the seventeenth century by Linnaeus and the other early systematists as they established the science of classifying the living world. A sort of hierarchy of similarity exists in nature, such that some organisms show more mutual similarity than do others. All backboned animals resemble each other more than they resemble any animals without backbones; but within this group all mammals are more similar to each other than any is, say, to a turtle or to a hagfish. And primates have more in common with each other than any primate has with a cow or an opossum. Darwin's most basic insight was to see that this self-evident hierarchy of resemblance among living things exists because they are all related genealogically. This is a magnificently simple explanation for the pattern of nature, and the only one from which one could predict how that pattern would express itself. Descent with modification: to Darwin this meant that all species, living and extinct, were descended from a single ancient ancestor, in a branching pattern that may be represented as we represent human genealogies: in the form of a tree, or a branching bush.

In *On the Origin of Species* Darwin, as befitted the most cautious and unconfrontational of men, was as circumspect as he could possibly have been about the implications of all this for the relationships by descent of humans—even though these implications were dazzlingly obvious. "Light will be thrown on the origin of man and his history" was all he would say at this point. But heat, as might have been expected, ruled the day as the obvious conclusion was drawn. The most celebrated of the acrimonious exchanges that took place in the months following the publication of the *Origin* was, of course, the encounter between Bishop Wilberforce and Thomas Henry Huxley, "Darwin's bulldog," at the meeting of the British Association for the Advancement of Science in June, 1860, when the issues of human relatedness to the great apes were brought out into the open. At the time the evolutionary view was widely caricatured as claiming human descent from the great apes themselves, which made marvellous newspaper and magazine copy; but of course the idea of common descent implies nothing of the kind. The apes, too, have become modified over the millions of years separating us from our common ancestor, which was neither a modern human nor a modern ape.

Huxley eloquently took the bull by the horns in 1863, when he published a small book of essays entitled *Evidence as to Man's Place in Nature.*" The thrust of this work was to demonstrate through comparative anatomy that the apes resembled humans much more than monkeys did. In one essay he described what was known of the history and the habits of the apes; in another he showed how embryology and anatomy demonstrate the affinities of humans to primates in general and to apes in particular (Darwin's notion of common descent providing the only plausible mechanism for this); and in the third he looked at fossils that might bear on human evolutionary history. These were limited to Engis (the adult specimen, which he correctly identified as a "fair average human skull, which might have belonged to a philosopher, or might have contained the thoughtless brains of a savage") and Neanderthal. He was impressed by the distinctiveness of the Neanderthal skullcap, but he concluded, largely because it had contained a modern-sized brain, that it could not "be regarded as the remains of a human being intermediate between Men and Apes." Although "the most pithecoid of known human skulls," it formed "the extreme term of a series leading gradually from it to the highest and best developed of human crania." Thus despite the advantage of an evolutionary perspective—from which he was at least able to ask whether "In still older strata do the fossilised bones of an Ape more anthropoid, or a Man more pithecoid, than any yet known await the researches of some unborn palaeontologist?"—Huxley saw in these specimens no more than evidence for great human antiquity. He therefore added nothing to what Schaaffhausen had already proposed—beyond, perhaps, establishing brain size as a vital criterion for

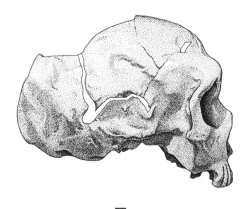
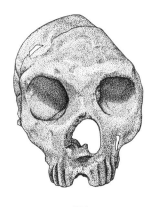

Side and front views of the Neanderthal skull from Forbes' Quarry, Gibraltar. Scales are 1 cm.   DM.

humanity. It is hard, however, to fault him for that: after all, the entire suite of distinctive material at his disposal consisted of one bizarre and highly incomplete specimen that was comparable to nothing else known to science.

   This did not stop others, however. The year after Huxley wrote, the geologist William King once more drew attention to the "remarkable absence" in the Neanderthal cranium "of those contours and proportions which prevail in the forehead of our species; and few can refuse to admit that the deficiency more closely approximates the Neanderthal fossil to the anthropoid apes than to *Homo sapiens.*" Indeed, King argued that the specimen was so apelike as to "lead one to doubt the propriety of *generically* placing it with Man," although "in the absence of the facial and basal bones" to advocate this view "would be clearly overstepping the limits of inductive reasoning." Nonetheless, in contemplating the Neanderthal skull King found himself compelled to conclude that the "thoughts and desires that once dwelt within it never soared beyond those of the brute"; and in a footnote, suppressing his desire to place it in a separate genus, he created for the specimen the new species *Homo neanderthalensis.* King's was the first formal recognition that another human species besides *Homo sapiens* had existed on Earth. And while its author was certainly bold in unhesitatingly arriving at this conclusion with just one fossil at hand, unbeknownst to him another specimen had already been found which demonstrated that the Neanderthal specimen was no aberration.

   The fossil skull from Gibraltar already mentioned had been found some time prior to 1848, when it was brought to the attention of the Gibraltar

Scientific Society. Unappreciated for what it was, it lay gathering dust in a small local museum until spotted by a visiting anthropologist in 1863. This gentleman arranged for it to be sent to London for examination by George Busk, Schaaffhausen's translator. The specimen itself clearly belongs to the human type represented by the probably male Neanderthal individual, though it is considerably more lightly built and may represent a female. The cranial vault is long and low, and the forehead slopes sharply back from the eyebrows, which are adorned with distinct ridges. The vault protrudes at the back and the face is large and somewhat projecting, with a wide nasal aperture and backward sweeping cheekbones. The significance of the Gibraltar skull was not lost on Busk, who immediately reported this proof that the Neanderthaler did "not represent . . . a mere individual peculiarity". The specimen was evidence that the Neanderthaler may indeed have represented a "race extending from the Rhine to the Pillars of Hercules." "Even Professor Mayer," Busk wrote, "will hardly suppose that a ricketty Cossack engaged in the campaign of 1814 had crept into a sealed fissure in the Rock of Gibraltar."

Alas! Busk's shrewd commentary fell on deaf ears. Huxley had taken to task Carter Blake, the secretary of the Anthropological Society in London, for suggesting that the Neanderthal bones were simply those of some poor idiot or hermit who happened to die in the cave. Yet pathology and idiocy (for certain inherited conditions that result in mental deficiency also involve abnormal development of the skull) continued to be favored explanations for the unfamiliar morphology of the Neanderthaler. This was especially true in Germany where the specimen, almost inevitably, became the center of an increasingly bitter controversy. The distinguished and antievolutionary pathologist and anthropologist Rudolf Virchow decided in 1872 that the Neanderthal individual was an aged man who had suffered rickets in childhood, head injuries in middle age, and chronic arthritis in his later years. These factors accounted for his odd morphology; and what's more, Virchow said, the fact that he was able to survive with these disabilities showed that he could not be ancient, for in a presettled society he could not have survived.

It was perhaps inevitable that the Neanderthal specimen, as the only truly distinctive early human then known, should become embroiled in the disputes over evolution itself in the years immediately following the publication of *On the Origin of Species*. But the variety of opinions that it elicited also reflected something else: the nascent science of paleoanthropology, as yet with only one fossil and with no well-established framework, was just beginning to feel its way. With no other comparable fossils, and no existing body of interpretation to provide such a structure, it is perhaps hardly surprising that almost every possible explanation for the odd appearance of the Neanderthaler was explored in the years immediately following its description. Disease, idiocy, trauma, the outer limit of normal variation, membership of a species totally distinct from modern humans—how many more possibilities could there be outside the realm of theology?

Well, one: the one, indeed, that most naturally occurs to (most of) us today. The Neanderthaler could have belonged to a species ancestral to humans, or to a collateral relative. But this idea was totally foreign to the received assumptions of the time, most of which dated back to a time well before the advent of darwinism. For years among anthropologists a major concern had been the origin of the human races. Some held that these had a single origin; others thought that they had been separately formed. While the biblical account of Creation held sway the idea of a single origin was naturally favored; but as the literal reading of Genesis lost ground in the later nineteenth century, the idea of multiple origins garnered more support. The Dutch science historian Bert Theunissen has pointed out that as darwinist ideas caught on, both these perceived possibilities of human racial origins were easily incorporated within an evolutionary context: either humans had diversified from one ancestral species, or each race was descended from an extinct nonhuman species. Emphasis was placed on the differences between human races, rather than upon their similarities, and typological notions—essentially of fixed form—ruled the day. This was, in fact, rather ironic, since even back in Buffon's time it had been permissible to speculate about the transformation of one infraspecific form into another, while it was now permissible only to talk in terms of transformation between species. But the upshot was that anthropologists could with little difficulty fit the Neanderthaler into the spectrum of defined human types as an added category. This was especially the case since European scientists were prone to rank the human species on a scale of perfection with themselves, not surprisingly, at the top. And since it was not uncommonly believed that the races at the top and bottom of the ladder were more dissimilar from each other than those at the bottom were from the apes, the Neanderthal type fit even more comfortably into this prevailing set of assumptions.

Another reason for neglecting the possibility that the Neanderthaler was a human ancestor or collateral relative was that in the 1860s the fossil record was not thought to be central, or even particularly relevant, to the problem of sorting out evolutionary relationships. The proof of evolution was sought and found in comparative anatomical and embryological similarities: to a greater or a lesser extent organisms share structural features and developmental sequences, and it is these that reflect and are explained by their common descent. The ebulliently darwinian German school of morphology, in particular, was based entirely on comparisons among living organisms to elucidate the order inherent in natural diversity; indeed, the dictum of the great German embryologist Ernst Haeckel, that "ontogeny recapitulates phylogeny," virtually made not only the fossil record but the comparative study of adult organisms irrelevant to the elucidation of evolutionary relationships. And although Haeckel's belief is not literally true that in its individual development each organism goes through all of the stages through which its antecedent species passed, it is nonetheless quite possible in principle to figure out relationships among extant animals without knowing anything of their fossil record. What fossils do, we now realize, is to add both a unique time dimension to our reconstructions of evolutionary history and to enlarge the comparative base available to us. But in the mid-nineteenth century the human fossil record, while accepted as evidence of human ancientness, was simply not associated with the idea of human descent.

Theunissen has drawn attention to one other widely held belief that militated against accepting the Neanderthaler as a precursor of modern humans. From the early years of the nineteenth century, the belief gained ground that modern Europeans had not developed in situ but were the descendants of people who had entered the region from somewhere in central Asia. Originally based on linguistic evidence of similarities between European languages and the more ancient Sanskrit of the Indian subcontinent, this idea gradually came to incorporate notions of race. By the third quarter of the nineteenth century this so-called Aryan hypothesis had been elaborated to the point that it envisioned a late Paleolithic or Neolithic invasion of Europe by agricultural peoples from the south and east. These "superior" invaders supplanted the indigenous hunting peoples, who were assumed to have been wiped out or otherwise to have become extinct. The Paleolithic Neanderthaler, then, could hardly have borne any direct relation to modern Europeans. Paleolithic Europeans were of incidental interest; human ancestry that mattered was to be sought in Asia. So while later finds, notably of two quite complete skeletons at the Belgian site of Spy in 1886, finally persuaded the world that here indeed was a distinct form of archaic human, few even then appreciated its significance. Indeed, the precise relationship of the Neanderthals to us remains debated to this day.

Even as the notion that, as Darwin put it, "species are the modified descendants of other species" was rapidly gaining favor among biologists, the antiquarians were transforming themselves into archaeologists and in the process establishing beyond doubt the ancientness of the human lineage. They were, moreover, moving beyond this to develop a chronology of the prehistoric human past. In 1865 the English archaeologist Sir John Lubbock published his *Prehistoric Times*, in which he adopted the scheme of successive Stone, Bronze, and Iron ages already proposed by two Danish scholars,

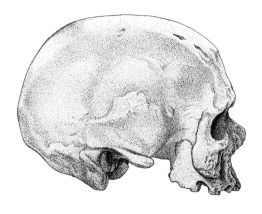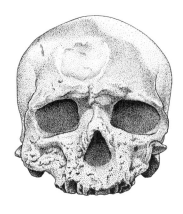

Side and front views of the "Old Man" cranium found in 1868 in the rockshelter of Cro-Magnon, southwestern France. Scales are 1 cm. DM.

Thomsen and Worsaae. Lubbock further subdivided the Stone Age into earlier and later periods: the Paleolithic, characterized by flaked or chipped stone tools, and the Neolithic, in which polished stone tools were used. It was already abundantly clear by 1865 that of all these periods the Paleolithic was by far the longest, and it was not long before it was in turn subdivided. Progress toward this subdivision surged in the middle 1860s, when the paleontologist Edouard Lartet and a local aristocrat, the Marquis de Vibraye, began to explore the archaeological potential of the caves and rockshelters in the limestone of the Dordogne region of western France. It had been known for years that shaped flints and bone were abundant in the earth filling such sites in the Dordogne, though Lartet's attention had initially been attracted by sites further to the south, such as the rockshelter at Aurignac, southwest of Toulouse, that in 1852 had yielded a large number of human burials in association with an extinct fauna and stone tools.

In 1868 Lartet and the English banker Henry Christy excavated a small rock shelter in Les Eyzies de Tayac, a village in the valley of the Vézère River which was destined rapidly to become dubbed the "Capital of Prehistory." At this site, Cro-Magnon ("Big Cliff" in the local patois), workmen digging fill for the adjacent railway line had discovered some human skeletons. Lartet and Christy's investigations revealed that at least five burials, one of an infant, had been made in the rock shelter. These people were of modern form but, as was by then regularly being found, were associated nonetheless with stone tools and the remains of extinct animals. It was Cro-

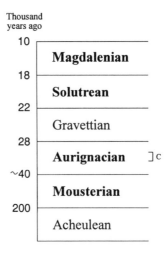

Thousand
years ago

10	
	Magdalenian
18	
	Solutrean
22	
	Gravettian
28	
	Aurignacian]c
~40	
	Mousterian
200	
	Acheulean

Chart showing the succession of cultures commonly recognized in the European Paleolithic. Names in boldface were coined by de Mortillet. Dates shown are approximate; in particular, the Aurignacian first occurred earlier in eastern Europe than the west (the eastern date is given here). The bracketed symbol "C" on the right-hand side of the chart gives the approximate duration of the Châtelperronian culture. DS.

Magnon that gave its name to the first modern people of the Dordogne region, and by extension of Europe as a whole.

Starting in 1865, the various parts of Lartet and Christy's great work *Reliquiae Aquitanicae* were published. In this landmark study Lartet suggested that although Le Moustier, Laugerie Haute, and La Madeleine, three of his sites near Les Eyzies, were clearly of the "age of simply worked stone without the accompaniment of domestic animals," they did not "possess a uniformity in the production of human industry." Clearly the Paleolithic had to be subdivided. As a paleontologist Lartet understandably tried to do this by means of the animals associated with the tools, recognizing a Cave Bear period, a Woolly Mammoth period, a Reindeer period, and so forth. Archaeologists, however, were unhappy with the idea of categorizing periods of human cultural development on the basis of zoological criteria, and it was not long before Gabriel de Mortillet reworked Lartet's chronology to correspond to the stone tool types that characterized the various periods.

De Mortillet's classification of 1872, enshrined in his great *Le Préhistorique* (1883), recognized four periods of distinctive stone tool making in the Paleolithic. Each of these was named after the locality in which it was first or

Fragment of reindeer bone from Le Chaffaud, France, showing engraved hinds. Probably of Magdalenian age, this piece was among the first works of Ice Age art to be recognized as such. Scale is 1 cm. DS.

best represented. The oldest of these industries, or cultures, was the Chellean (later changed to Abbevillian), named for one of Boucher de Perthes's sites in the Somme River valley. Flint tools of this age included massive handaxes made by knocking flakes off a "core" until a standard shape was obtained. This was followed by the Mousterian, in which tools were fashioned on large flakes detached from a core that had been shaped to predetermine their form. Next came the Solutrean, characterized by incredibly finely worked "laurel-leaf" points. The final period of the Paleolithic was the Magdalenian, when fewer tools were made of stone and more of organic materials such as bone and antler. Later editions of de Mortillet's work added the Aurignacian period between the Mousterian and Solutrean, as the first period of the Upper Paleolithic, and the Acheulean following the Chellean. In its essentials, this chronology of the Paleolithic survives today. Despite the general antipathy to Darwin's ideas that reigned in a France still in thrall to the shade of Cuvier, de Mortillet was also an enthusiastic darwinian and was an early supporter of the idea that the roots of *Homo* were ancient indeed. He it was who coined the term "eoliths" (dawn stones) for the simple tools that he expected "Homo-simius," the human precursor, to have made.

As the complexities of the archaeological record of the Paleolithic were emerging, the earliest evidence of artistic activity by Aurignacian and later peoples was coming to light. As early as 1833 a baton and harpoon made of antler, both decorated with engraving, were found in the cave of Veyrier in Switzerland, and at around the same time an engraving of two deer on a plaque of reindeer bone was recovered from the French cave of Chaffaud. Thought by its discoverers to be Celtic, the Chaffaud piece was recognized by Lartet to be of Paleolithic origin, and he published it as such in 1861, along with an engraving of a bear's head which had been excavated in the cave of Massat in the foothills of the French Pyrenees. At Aurignac, too, engravings on bone had been found; and in 1864 Lartet and Christy published a detailed discussion of this evidence for ancient artistic activity. Objectively, there could be little doubt about the ancientness of such pieces. After all, most of them had been found in situ, buried under thick piles of sediment and in association with extinct animals whose representations they sometimes bore. Nonetheless, the idea of prehistoric art took some time to be absorbed by the establishment. But by 1867 the veracity of Paleolithic "portable" art was sufficiently well established that fifty-odd examples of it were placed on exhibit in the great Universal Exposition, held in Paris that year.

Cave art from the Paleolithic took much longer to be accepted, however. The first such discovery to be made was that of the spectacular painted ceiling of the cave of Altamira, in northern Spain, by the young daughter of Don Marcelino Sanz de Sautuola. As her father excavated in the floor of the cave in 1879, searching for prehistoric artifacts, she (who alone could stand upright beneath the low ceiling) looked up and saw in the lantern-light the now-famous polychrome representations of bison, horses, and other large mammals. Her father recognized the similarities between these images and the engravings with which he was already familiar from portable art, and ultimately he concluded that the paintings were indeed Paleolithic. Initial reaction to this remarkable find was favorable, but in academic circles a reaction soon set in, and Altamira was condemned as a fake by one prehistorian after another. Supporters of de Sautuola were soon in a tiny minority. It was not until corroborating finds began to be made at other sites near the end of the century that opinion turned in the now-dead de Sautuola's favor, and the cave art of the Upper Paleolithic was finally accepted.

While de Mortillet was imagining the nature of the tools that might have been produced by an ancient human precursor (and virulently attacking the authenticity of Altamira), others were conjecturing the attributes of the creature itself. In the forefront of these was the German zoologist Ernst Haeckel. One of Darwin's most enthusiastic supporters, Haeckel was quite ready to plunge into areas such as the differentiation of different human types, where the master himself trod gingerly if at all. In his long and highly popular 1868 book *The History of Creation* Haeckel described and illustrated a highly specific tree of life that ran in twenty-two stages from spontaneously generated "formless matter" at the root to humanity at the top. The apes occupied little twigs near the pinnacle, but on the trunk beneath mankind lay "Ape-Men." The living apes, Haeckel pointed out, cannot be regarded as ancestral forms to man, but an ancestor there had to be. This "hypothetical primaeval man . . . who developed out of the anthropoid apes" had a skull which was probably "very long, with slanting teeth . . . the hair covering the whole body was probably thicker [than in modern humans] . . . their arms [were] comparatively longer and stronger; their legs, on the other hand, knock-kneed, shorter and thinner . . . their walk but half erect." What's more, this rather unattractive beast would have lacked the feature that Haeckel believed set off humanity most clearly from its closest relatives in nature: articulate speech. The human evolutionary stage of speechless ape-men Haeckel called "Alalus, or Pithecanthropus"—a name shortly to become famous in another context.

In his own equally lengthy but more readable contribution *The Descent of Man* (1871), Darwin was a little less exuberant; indeed, despite his title, he was more occupied by the question of sexual selection (female choice) as a factor in evolution than by the consideration of the human descent per se. The fossil record he ignored almost entirely, merely remarking in passing that the Neanderthal skull was "well developed and capacious." He had no difficulty, however, in concluding that the anatomical and embryological similarities shared by humans and apes were far too numerous to be due to anything but common descent from an ancestor more generalized than any of its living descendants. This ancestor was, according to Darwin, a "hairy quadruped, furnished with a tail and pointed ears, probably arboreal in its habits, and an inhabitant of the Old World."

Darwin's argument was a biological one, structured as we've seen around the use of embryological and anatomical comparisons to demonstrate the commonality of descent between human beings and other living things. It is thus hardly surprising that in his consideration of human descent he did not look to archaeological information to flesh out the picture of our species' past. De Mortillet had no such reluctance, however. He believed that cultural evolution was intimately tied in with the later part of the history of human divergence from "lower" forms. He gave the name *Anthropopithecus* to a hypothetical human ancestor that had lived late in the Tertiary Period. No fossils of such an ancestor intermediate between apes and humans were known, but to de Mortillet a graded series had to have existed between an ape-like ancestor and modern humans. In his later works he came to see the Neanderthals, by then associated with the "Mousterian" culture, as a stage in this process, still bearing some hints of ape ancestry. And just as he imagined that in the past a continuous succession had existed of steadily more human species, he perceived a constant linear sequence of cultures throughout the Paleolithic, each one giving rise to the next. And although the concept that the Neanderthal fossil represented a form ancestral to modern humans was the one theoretical possibility that was not entertained by any of the early biological interpreters of the specimen, the archaeologist de Mortillet had hit on a theme which was to exercise an abiding fascination upon paleoanthropologists, one that runs deep in their science to this very day.

3

Pithecanthropus

During the 1870s and 1880s the Paleolithic archaeological record continued to accumulate much more rapidly than the human fossil record. An occasional modern-looking human skull turned up at this or that ancient archaeological site; but apart from the discovery in 1886 of two nearly complete Neanderthal-like adult skeletons in the Belgian cave of Spy, no new archaic human fossils were discovered during this period. The Spy finds were of the greatest importance both because they confirmed that the original Neanderthaler was no aberration and because they were found in an undisturbed archaeological context that established beyond doubt both the great antiquity and the "Mousterian" cultural association of these early humans. The describers of the Spy specimens, Julien Fraipont and Max Lohest, considered that these humans were much "advanced" beyond the apes and must have been descended from forms more apelike yet; but despite this evolutionary viewpoint they ultimately concluded that they represented a "human race." Thus it was that the 1890s arrived with no accepted physical evidence for the remote "apish" human ancestor whose existence more and more scientists were willing to hypothesize.

One of the many reasons for the sparseness of the nineteenth-century human fossil record is that all the early human remains that composed it were incidental finds. They had turned up as by-products of quarrying or construction, of fieldwork by geologists, or of the excavations of antiquarians and prehistorians in search of artifacts. Despite the paleontological expertise of Edouard Lartet, who had done as much as anyone to establish human antiquity, no ancient human fossil had ever been found as the result of fossil collecting in the paleontological tradition. Virtually every ancient human specimen then known had been recovered from sediments in cave

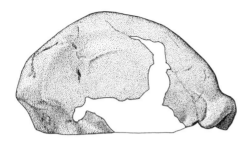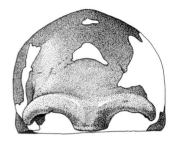

Side and front views of one of the Neanderthal crania (Spy 1) from the Belgian site of Spy. Scales are 1 cm.   DM.

mouths; none had come from the layers of sedimentary rocks in which fossil bones are normally found. Few cave sites contain deposits that are more than several tens of thousands of years old; only in the strata that compose the general landscape are truly ancient fossils to be found. And to find truly ancient human fossils in the scientific climate of the late nineteenth century required a certain kind of inspiration.

Nobody knows exactly why Eugene Dubois, a young Dutch anatomist and physician, should in 1887 have sailed for the Dutch East Indies with the avowed intention of finding the remains of fossil man—although he certainly seems to have been fired up by a lecture he had heard in medical school from the charismatic Ernst Haeckel (who had "predicted" the existence of a fossil ape-man and had urged exploration for it in the "bone caves" of the Malay archipelago). Dubois, though a keen naturalist from childhood, had had a relatively conventional education in biology and anatomy—an education, what's more, in the Germanic tradition, which tended more perhaps than any other to eschew the fossil record in documenting evolution. It is known that Dubois was uncomfortable in his position as an instructor of anatomy at the University of Amsterdam, and his reasons for trying his luck at finding human fossils in the East Indies rather than elsewhere are clear enough; but there is nothing known about his background that even comes close to explaining why he should have abandoned everything—his job, his prospects as an anatomist—to become the first human paleontologist, especially since at the time his mother country had hardly anything in the way of a paleontological tradition.

Nonetheless, at some point Dubois became a convinced darwinian, and one of those who rejected the Haeckelian notion that fossil evidence for human evolution was not essential (despite Haeckel's prediction that it would be found). Dubois felt, deeply, that the course of human evolution could be demonstrated and described only through fossils, and he conceived a mission to find them. In view of this it seems perhaps odd that Dubois's view of the Neanderthal skull and others like it was utterly conventional. He saw no hint in these fossils that they might be precursors to modern humans; even if these specimens were not pathologically deformed, he thought, they showed only that a primitive race of human beings had existed. But in this conventional view we can see a key to Dubois's unconventional actions: Neanderthals could not be prehumans because, as received wisdom indicated, people had arrived in Europe only after becoming fully human. Clearly, then, to find the precursors of *Homo sapiens* it was necessary to look outside Europe.

Where? Darwin's candidate as the cradle of mankind had been Africa, tropical home of two of the three great apes, the chimpanzee and the gorilla. Dubois had no private means, however, and wherever he was to go he needed employment: which is why he eventually signed on as a medical officer in the Royal Dutch East Indies Army. This action provided him with a means of getting at government expense to the region inhabited by the third great ape, the orangutan. And the orangutan had figured largely in many early theories about human origins and humanity's place in nature. Dubois's reasons for choosing the Dutch East Indies as the sphere of his operations were thus not solely financial. These islands were tropical, and it was generally agreed that the human ancestor lived in the tropics. Their extinct fauna, insofar as it was known, was thought to resemble that of India, where as early as the 1870s fossil apes had begun to be found in sediments associated with the building of the Himalayas. In Dubois's day this material consisted of parts of an upper jaw of a form then known as *Palaeopithecus sivalensis*, thought by its describer Richard Lydekker to be a relative of the chimpanzee; in 1886 a tooth indistinguishable from an orangutan molar was also found. And the East Indies were, of course, the home of the orangutan, the third living great ape, as well as of some of the lesser apes, the gibbons. If humans were descended from apes, then a part of the world with both living and, presumptively, fossil apes offered enticing prospects for finding a human ancestor. And from the point of view of discovery the calcareous rocks of Sumatra and other East Indian islands were known to be riddled with caves—just like the sites in which the European fossil humans had been found.

Still, Dubois's move was a huge gamble, and it is truly remarkable that it paid off. At first it didn't, although his initial efforts at cave exploration in Sumatra were rewarded with an impressive collection of fossil bones of living species on the basis of which he was able to gain government support for his work. But continuing cave explorations were disappointing, and in

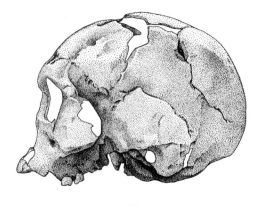
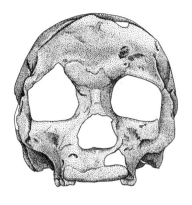

Side and front views of the fossil modern human skull found at Wadjak, Java. Scales are 1 cm.   DM.

1890 Dubois transferred his attentions to Java, where a fossilized human skull had been found in a rockshelter at Wadjak, in the eastern part of the island. The skull was that of a modern human, although Dubois concluded that it differed sufficiently from those of the current inhabitants of Java to represent a distinct earlier population. Cave investigations continued to be unexciting, however, so Dubois directed his workforce—a gang of convicts under the supervision of two army corporals—to excavate in open-air geological deposits already known to yield mammal fossils, often those of extinct species. In November 1890 he found his first human fossil bone, among those of many other vertebrates, at a site called Kedung Brubus. The specimen was an unimpressive fragment of lower jaw, with only one premolar tooth and the socket of a canine. Dubois wasn't able to make much of it, and neither has anyone since. But before long, better material was at hand.

In 1891 Dubois began to investigate some alternating beds of sandstone and volcanic rocks near the village of Trinil, on the banks of the Solo River in central Java. Unlike conventional paleontologists, who—except when collecting very tiny fossils—generally look for bones that are eroding out of exposed sediments and excavate only to recover specific fossils, Dubois's crew proceeded like archaeologists, digging pits in the ground. At Trinil the convicts dug a vast pit, forty feet across and eventually fifty deep. The many fossils found were packed in teak leaves and dispatched periodically to Dubois, who spent most of his time at his base in Tulungagang, a considerable distance away. Unfortunately, this approach did not permit pre-

Front and side views of the original Trinil Pithecanthropus *skullcap discovered by Eugene Dubois in 1891. Type specimen of* Homo erectus. *Scales are 1 cm.*   DM.

cise localization of where each fossil was found—a problem that came back to haunt Dubois, and continued to bedevil paleoanthropology in Java long after his time. But the digging was effective: in September 1891 a hominoid molar tooth was found (Hominoidea is the name of the zoological super-family that contains man and the apes), and in the next month a skullcap emerged. This was broadly equivalent in completeness to the original Neanderthal specimen, though its braincase was much less capacious. Dubois first reported these remains as fossil chimpanzees, having concluded that they resembled neither gorillas nor orangutans. But he soon changed his mind.

The next year, after the floods of the rainy season had subsided, workmen found a femur in an extension trench at Trinil. It was reportedly recovered from the same stratigraphic level as the skullcap and tooth, but some distance away—variously reported at different times as fifteen, twelve, or ten meters. Though it bore a large pathological excrescence this femur was just like that of a modern human, and it indicated that its possessor had walked erect. Since a reappraisal of the cranial vault had shown it to be too capacious for any living ape, Dubois concluded that what he had before him were the remains of an apelike man rather than a manlike ape. In homage to Haeckel he gave it the name *Pithecanthropus erectus*, "upright ape-man." Haeckel returned the compliment by wiring Dubois "congratulations to the discoverer of *Pithecanthropus* from its inventor."

Rather as in the Neanderthal individual, the eyes of *Pithecanthropus* were overhung with ridges. The vault of the somewhat smaller skull was long and low (yet higher than that of an ape) and sharply angled at the rear. The braincase was composed of thick, heavy bone. Dubois estimated that the individual's brain had been around a thousand milliliters (1,000 ml) in volume. This is at the extreme lower limit for modern humans, who average about 1,400 ml (as do Neanderthals, but Dubois declined to compare his specimen with Neanderthal and Spy because he thought them pathological as well as variants of modern humans). In later years Dubois dropped his estimate of the brain volume of the Trinil skullcap to 900 ml; modern calculations place it at 940 ml. For comparison, chimpanzee brains average around 400 ml and those of the much larger gorilla, about 460 ml. Brain size is, of course, to some extent dependent on body size, but Dubois figured on the basis of the femur that the stature of his early human had been similar to that of modern Europeans. As to the age of his *Pithecanthropus*, Dubois conjectured from the huge associated fauna that the specimens dated from the late Pliocene or early Pleistocene, intermediate between the Indian Siwalik fauna and the human specimens known from the late Pleistocene of Europe.

Was this upright ape-man the human ancestor he had been seeking? In a lengthy description published in 1894, Dubois argued that it was impractical to classify *Pithecanthropus* either in the human genus *Homo* or as an ape. Instead it stood between them, as the notion of evolution suggested some creature must have done. He proposed that a gibbonlike ancestor had given rise to the extinct Indian "chimpanzee," which had in turn evolved into *Pithecanthropus erectus*, whence *Homo sapiens* (including the Neanderthals) had sprung. This suggested that major anatomical shifts had occurred between each of these successive stages, but Dubois saw no problems with jumps of this kind. Furthermore, his new species suggested to him that Lamarck and Darwin had been right to conclude that an upright bipedal way of walking had been the first new adaptation to characterize the lineage leading to human beings. Only later had such traits as manual dexterity and a large brain been acquired. But while the femur showed undoubted evidence of human bipedality, the skullcap was different, retaining vestiges of the "ape-type" in its skull; thus, although *Pithecanthropus* had certainly advanced some way toward human status, it could not yet be classified in Hominidae, the family of mankind. And while he finally decided to place it in its own family, Pithecanthropidae, Dubois continually returned in his description to comparisons with the chimpanzee and the gibbon. These latter "lesser" apes brachiate through the trees, swinging arm over elongated arm. But on their rare descents to the ground, they walk bipedally. This observation in particular deeply impressed Dubois, leading him down a blind alley into which many others followed.

Such complexities notwithstanding, the message that Dubois took with him to Europe when he returned there in the summer of 1895 was a simple one: during the late Pliocene or early Pleistocene, an upright being intermediate between ape and man had lived in Java. Parenthetically, one might note that in those days nobody knew exactly how long ago that was; Lyell had estimated in 1863 that the Pleistocene had lasted 800,000 years, while by 1900 another English geologist, W. J. Sollas, had reduced it to half this.

Despite the fact that Dubois's 1894 description of his *Pithecanthropus* fossils had been published in remote Batavia (now Jakarta), and in a special publication rather than in a recognized journal, news of his find preceded his return to Holland. Most scientists' initial reaction was to question the association between the skullcap and the femur. Some critics dismissed the cranium as that of an ape (possibly, according to some authorities, a giant gibbon), while attributing the femur to a modern human. Others, particularly in England and Ireland, were willing to accept the skullcap as human, although of a rather low order. The Dublin anatomist Daniel Cunningham, for one, was prepared to see *Pithecanthropus* as a form which had given rise, via the Neanderthals, to modern humans, though he regarded the femur as fully modern. And in America the famous paleontologist O. C. Marsh fully concurred with Dubois's conclusions. At the beginning, however, the poorly established physical association between the femur and the skullcap was seen as undermining Dubois's interpretation.

Once again, we have to bear in mind that this debate was conducted in the absence of any substantial body of comparative material. The idea that the Neanderthals were not simply an odd ancient race of modern humans was still foreign to most points of view, which meant that the *Pithecanthropus* fossils were all that stood between humans and apes. And to many the skullcap and femur appeared to tell different stories, so that to exploit the doubt over their association was also to simplify their interpretation. As important, in the closing years of the nineteenth century there was no coherent body of expectation against which the fossil record could be tested. Agreement was reasonably complete that descent with modification was indeed the explanation for the diversity of living organisms, but opinions varied greatly on how evolution actually worked. Darwin's mechanism of natural selection had its adherents, but by no means all biologists subscribed to it. Besides, despite having called his book *On the Origin of Species by Natural Selection*, Darwin had barely considered the central question raised in its title. Natural selection dealt with the morphological transformation of lineages, which depending on your point of view is not quite the same thing as the origin of species, or not the same thing at all. Even such staunch supporters of evolution as Thomas Henry Huxley had taken Darwin to task for concentrating so single-mindedly on the slow accretion of small changes. In reviewing Darwin's book Huxley wrote that "Mr Darwin's position might . . . have been even stronger than it is, if he had not embarrassed himself with the aphorism 'Natura non facit saltum' [Nature makes no leaps] . . . nature does make jumps now and then, and a recognition of the fact is of no small importance." Huxley echoed the view of many, while still others clung to Lamarckian ideas and the notion that evolution was somehow goal-directed. Small wonder, then, that with theory providing no consistent guide to the interpretation of the scantiest of fossil records, wildly varying explanations of Dubois's *Pithecanthropus* multiplied rapidly.

It has often been reported that these interpretations were generally unfavorable to Dubois's point of view, and that it was this negative reaction that caused Dubois to withdraw from the fray, first himself, and later his fossils as well, in the years after 1900. Recently, however, Bert Theunissen has convincingly shown that Dubois was much more successful than usually supposed in winning converts to the idea that *Pithecanthropus* was an ancestral human form, or one that was at least close to human ancestry. Between 1895 and 1900 Dubois had traveled widely in Europe, showing off his specimens at various international congresses. Many of his colleagues turned out to be prepared to modify their ideas on viewing the original specimens. Most eventually accepted that the femur and the skullcap of *Pithecanthropus* were indeed associated; and virtually all agreed, upon seeing how thoroughly the bones were fossilized, that they were very old indeed. Not that opposition to Dubois's interpretation of his fossils was lacking; Virchow, for one, loudly proclaimed that the remains were those of a giant gibbon, though this interpretation may have owed less to scientific considerations than to his vicious quarrel with Haeckel, who supported Dubois with equal energy. But even those who saw in *Pithecanthropus* a more apelike or a more humanlike form than Dubois did, at least agreed that here was evidence of a manlike ape or an apelike man. And Dubois, in his turn, found support for his theories in the fact that both extremes of opinion were represented. A few authorities held out the possibility that the specimen was pathological, or that it merely represented an extinct human race; but they were in a small minority.

Given the stir created by the new material from Java, it was inevitable that the Neanderthals should once again come under scrutiny. And although Dubois never shifted from the idea that these were simply an extinct race of *Homo sapiens*, other scientists who accepted that *Pithecanthropus* lay on or close to the line of human descent now found it easier to envision the Neanderthals as a later stage in the same process. Influential among them was the German anatomist Gustav Schwalbe, who produced long monographs on both *Pithecanthropus* and Neanderthal in the years around the turn of the century. Schwalbe perceived a morphological series from the apes to *Pithecanthropus* to the Neanderthals (which, ignoring King's earlier name, he christened *Homo primigenius*) to *Homo sapiens*. He considered the Neanderthals closer to *Pithecanthropus* than to modern humans. As his ideas developed, Schwalbe came to see this series as a line of descent, reflecting a consensus that was gathering strength, particularly in England. In the end, Schwalbe's minutely documented analyses played a critical role in getting the Neanderthals accepted as a distinct form of early human: an acceptance that was also helped along by the decline of the Aryan theory in the closing years of the nineteenth century and by the gradual disappearance of the single- versus multiple-origin dispute in the discussion of human races.

But Schwalbe's intense interest in *Pithecanthropus* seems to have had a more direct and unintended effect on Dubois, who appears to have sensed his intellectual capital disappearing as detailed descriptions of his fossils came out under the signatures of other scientists. Eventually he withdrew the bones from scientific access (they remained locked away until 1923). During this period he devoted himself to the studies of comparative brain size in mammals that he had already begun with the aim of showing that *Pithecanthropus* indeed stood intermediate between the apes and mankind. Adhering to a notion of evolution by abrupt change and of an inner urge to "perfectionism," as opposed to the gradual change in response to external factors postulated by Darwin, Dubois concluded by late in his career that mammalian brain evolution had occurred via a series of sudden doublings in size, each due to an additional division of the primordial brain cells. These had proceeded such that, compared to body volume, apes (including gibbons) had one fourth the amount of brain that humans have; the majority of mammals ("Ruminants, Cats, Dogs, etc."), one eighth; rabbits, fruit bats, and various other forms one sixteenth; mice, moles, and leaf-nosed bats one thirty-second; and shrews and "common Small Bats," representing the most primitive mammals, one sixty-fourth. In this series, one stage was glaringly missing: the stage between the apes and man, with a brain one-half the relative size of the human one. It was in this position that a form transitional between the apes and man would have to fit. But while the femur seemed to indicate that *Pithecanthropus* had been as large as modern humans, the skullcap had held a brain not one half but about two-thirds modern size. There was a way out, though: if *Pithecanthropus* had boasted gibbonlike body proportions, with relatively short legs and a large thorax, it would have been much heavier than a human with a femur of similar length. Using the greater body weight which such proportions would yield, the brain to body ratio of his "Java Man" fell neatly into the doubling series.

It was thus in defense of his original conclusion that *Pithecanthropus* was transitional between ape and mankind that Dubois claimed gibbonlike characteristics for it. But the ape ancestor thus became not a great ape but a lesser one; and as Dubois grew older, the gibbon hypothesis took on a life of its own. By 1935, for example, Dubois felt able to assert that *Pithecanthropus*, although an upright walker, had upper limbs that "still exerted locomotor functions . . . in a similar manner as in the Gibbons"—which, as we've seen, are highly specialized arboreal armswingers. By this time Dubois was thoroughly out of step not only with evolutionary theory and growing knowledge of brain structure and function, but also with advancing interpretations of the human fossil record. After about 1900, then, he contributed little to the mainstream of debate on human evolutionary history—even though he himself had done more than anyone else to spur that debate. Fixating on his *Pithecanthropus*, he refused to acknowledge the significance of other near-human fossils that were coming to light.

4

The Early Twentieth Century

In 1900 three investigators independently rediscovered the principles of inheritance first enunciated thirty-four years earlier in an obscure journal by the Moravian cleric Gregor Mendel. Concepts of inheritance up to that time had been vague and sometimes Lamarckian. Most of them involved some idea of "blending," since it was widely recognized that in sexually reproducing species offspring, while not identical to either parent, tend in many ways to resemble them both. It was Mendel's great insight to study the transmission of traits that did not behave this way. Growing flowering peas in his monastery garden, he carried out breeding experiments on plants that bred true for traits that existed in two alternative forms: stature (tall or short), seed color (green or yellow), and so forth. Mendel elegantly showed that such traits are specified by inherited factors (which we now know as genes) that exist in a variety of forms (alleles). Each individual receives one allele from each parent. One allele of the pair each individual thus possesses may be dominant over the other (recessive) allele, which is not expressed in its presence. In Mendel's peas, for example, the "tall" allele was dominant over the "short" one, so that an individual with one of each would still be tall. Nonetheless, two potentially different alleles are still present in each individual, and each has an equal chance of being passed along to the next generation. What's more, Mendel found that each discrete trait (stature, seed color) might turn up in combination with any other; each was transmitted independently.

Today we know that genes behave this way because each individual possesses a double set of chromosomes, one inherited from each parent, and along which the genes are arranged sequentially. Each chromosome of a pair carries one allele. The traits Mendel chose to look at were all carried on different chromosomes, and each was determined by only one gene. He was lucky. In most cases, as suggested by the fact that offspring tend overall to look a bit like both parents, most physical characters are influenced by more than one gene—which usually occur on different chromosomes—while most genes affect more than one character.

Mendel's acute observations, published in an obscure local journal, languished unappreciated for over three decades during which other notable advances in the science of inheritance were nonetheless made, notably the anti-Lamarckian demonstration by the German biologist August Weissmann that there was a total separation between modifications of the physical being and of the material of inheritance. But the rediscovery of Mendel's principles unleashed an astonishing burst of activity in the nascent field of genetics. If inheritance was discrete—and even if multiple genes were usually involved—the patterns of heritability could be modeled mathematically, and the study of inheritance could finally be put on a scientific footing. And if evolution was the sum total of perturbations in the transmission of genetic information between generations, here was a mechanism by which evolutionary phenomena could—indeed, must—be explained. So, even though Darwin and his fellow naturalists such as Wallace had been perfectly able to identify and describe the basic phenomena of evolution in the total absence of an accurate notion of how inheritance worked, genetics suddenly appeared to many to be the key to unraveling the mysteries of natural diversity. And although it took several decades for the systematists, paleontologists and geneticists to come together on this issue, when they did so it was to have profound implications for the future of evolutionary biology.

Paradoxically, though, just about the first thing a budding paleoanthropologist is taught nowadays in courses on evolutionary change is the so-called Hardy-Weinberg Principle. First adumbrated in the early years of this century, this demonstrates mathematically that the distributions (otherwise known as frequencies) of alleles within a freely interbreeding population of organisms should remain more or less constant from one generation to the next. It is, then, a demonstration of stability, not of changeability, and in later decades it became backed up with a great deal of literature on the subject of genetic homeostasis: the inertial tendency of populations not to change. And it became the starting point of the science that we know nowadays as population genetics, whose practitioners usually regard themselves as evolutionary biologists. But evolution, of course, involves change; and the big question is, how does such change occur? Translated to the terms of population genetics, Darwin's ideas of natural selection involve gradual shifts in gene frequencies within populations, as favorable genes or gene combinations spread from generation to generation at the expense of less favorable ones. But as people like dog breeders, selecting for longer legs or shorter muzzles or whatever, had known before evolutionary ideas had ever taken shape, by simply shifting gene frequencies you could produce neither true anatomical innovation—you were restricted to producing variations on existing themes—not new species. Yet evolution depends both on anatomical change and on the production of new species, and at the end of the nineteenth century the relationship between these two different things remained far from clear—as, to be quite frank, it still does.

To begin with species, as early as 1865 the French naturalist Pierre Trémaux had lamented that "of definitions of species there are as many as there are naturalists," and this is not a situation that has materially changed in the years since. And Trémaux also put his finger on the crux of the matter: "Two principal notions have served to define the species: resemblance between individuals and ability to reproduce. [T]he first of these conditions . . . is . . . the consequence of the second." This comment holds with equal force today: species are still viewed by the majority of biologists as reproductive communities composed of individuals who more or less resemble each other and who are at least potentially able to interbreed among themselves while being unable to do so with members of other such communities. Trémaux was quite right to point out that it is because of their common genetic heritage that the individuals composing a species look similar; nonetheless, it is almost invariably on the secondary criterion of similarity that we recognize individuals as belonging to the same species. Even among living organisms, let alone among fossils, there are vast problems both in principle and in practice of recognizing exclusive or inclusive reproductive communities, and it is essentially because of these problems that there are still as many definitions as naturalists. But while at least in theory we can recognize species from some aspect of resemblance when we look around us at the living world (though, remember, Darwin himself had had trouble with this), under strict darwinian tenets we can't do that in the fossil record unless we take a single slice of time. For, through slow, gradual change under the guiding hand of natural selection, darwinian species are expected to transform themselves out of existence. Reproductive continuity stays unbroken over the millennia, but anatomies nonetheless change, eventually greatly.

The alternative view, popular during the late nineteenth century when natural selection was not greatly in vogue, is that species maintain their identities in time as well as in space. If we accept this, we have to be able to explain how one species is able to transform itself into another in a rather short space of time—one so short that the likelihood of our picking it up in the fossil record is negligible. One way of doing this seemed almost self-evident from the work of the geneticists. It had long been known that while most individuals in a population differed from each other only in minor ways, occasionally a "sport" might pop up that was quite different from the norm. Following his rediscovery of the "particulate inheritance" of Mendelian genetics, the Dutch botanist Hugo de Vries showed that spontaneous changes, or "mutations," in an allele could account for innovations of this kind. Taking this idea further, he proposed in 1901 that new species arose in a similar manner, by sudden steps involving change in one major character or set of characters—an idea of speciation that was rapidly endorsed by many other influential geneticists. What's more, the Mendelians rapidly became more or less unanimous in identifying mutation pressure—the rate at which mutations occur—as the driving force behind evolutionary change.

This new influence tended to distract attention from other considerations relating to the origin of species; and it also set the scene for a rift to develop between the geneticists and the more traditional naturalists, systematists, and others. The geneticists, while rapidly developing a body of theory to cope with the obvious fact that inheritance was in most cases more complex than Mendel's simple rules suggested, nonetheless concentrated on the role in evolutionary innovation of discontinuous characters (for example, stature in peas). The naturalists, on the other hand, many of whom were becoming more inclined to accept natural selection, emphasized the importance of continuously variable characters (such as stature in humans) in the evolutionary process. The paleontologists did not contribute materially to this debate, in which they simply tended to take one side or the other: among the paleoanthropologists, for instance, Dubois supported the mutationist viewpoint of his countryman de Vries, while many others inclined to gradualist interpretations of evolutionary change.

It was in this rather unsettled climate of evolutionary opinion that the human fossil record continued gradually to accrete in the early years of our own century. New finds continued to be made by archaeologists or by accident, for few scientists followed in Dubois's footsteps and actually went in search of human fossils, especially after a German expedition under the leadership of Margarethe Selenka spent 1907–1908 at Trinil without finding any additional human specimens. This expedition was not without its successes, though, for it did collect many nonhuman vertebrate fossils. Dubois himself had paid little attention to his own enormous assemblage of such material following his return to Europe; indeed, his collection remains incompletely studied to this day. So it was particularly important that analysis of the Selenka materials confirmed the age of the Trinil fauna as early or middle Pleistocene, even though this complicated the picture somewhat. At that point the Neanderthals were considered to be of middle Pleistocene

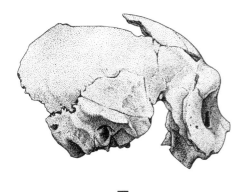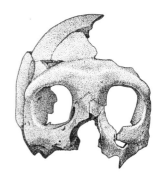

—   —

Front and side views of Krapina cranium C, one of many fragmentary Neanderthal fossils found at that Croatian site at the turn of the century. Scales are 1 cm.   DM.

age, antedating as they did the Cro-Magnon types of the late Pleistocene; and if both Neanderthals and *Pithecanthropus* were of Middle Pleistocene age, how could the latter be ancestral to the former? This puzzle was to go unsolved for some time, even as the view that *Pithecanthropus* and the Neanderthals stood in the direct line of human descent, or at least represented stages in that line, was gaining credibility.

Not, however, for long. Between 1908 and 1911 the French cave sites of La Chapelle-aux-Saints, Le Moustier, La Ferrassie, and La Quina all yielded up complete or multiple skeletons of Neanderthals which at last provided an opportunity to appraise the entire bony anatomy of this early human form. These complemented a series of less complete Neanderthal remains that had been excavated in the cave of Krapina in Yugoslavia between 1899 and 1905, a fragmentary braincase and other materials found at the German site of Ehringsdorf in 1908, and various other bits and pieces found over the first quarter of the century. But for a variety of reasons it was the skeleton of the "old man" of La Chapelle that, through the voluminous publications of the influential French paleontologist Marcellin Boule, became the Neanderthal archetype. And Boule was dead set against viewing the Neanderthals as anything but an offshoot of the human lineage that had died out without issue. Anatomically, Boule claimed, the Neanderthals had possessed divergent big toes (hence grasping feet, on which, moreover, weight was borne, apelike, on the outer edges), a slouching posture, bent knees, short and thick necks, and inferior brains. What's more, Boule believed that modern human forms already existed at the time of the Neanderthals, and he also pointed out that in France the Mousterian (or Middle

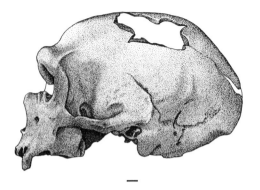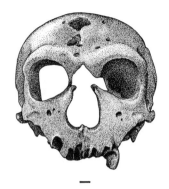

Front and side views of the cranium of the "Old Man" of La Chapelle-aux-Saints, Corrèze, France. Scales are 1 cm. DM.

Paleolithic) stone tool-making tradition of the Neanderthals had been rather abruptly replaced by the Upper Paleolithic tool assemblages associated with modern human types. This suggested to him that the Upper Paleolithic had been developing elsewhere for quite some time, and he took this as further evidence that the Neanderthals were but a side branch in human evolution. Boule was quite happy to accept the Neanderthals within the genus *Homo*, as *Homo neanderthalensis*; but, he wrote of them in 1913, in the final part of his long monograph on the La Chapelle-aux-Saints skeleton, "what a contrast with the . . . Cro-Magnons, [who with their more elegant] bodies, finer heads, large and upright foreheads . . . manual dexterity . . . inventive spirit . . . artistic and religious sensibilities . . . [and] capacities for abstract thought were the first to deserve the glorious title of *Homo sapiens*!"

One is supposed not to write Whiggish history, in which figures of the past are measured by their importance in contributing to today's orthodoxies (and indeed, my own opinions coincide with some of Boule's views). But I have to confess that, try as I may, in reading Boule's 1911-1913 monograph I cannot avoid total incredulity at his cavalier treatment of *Pithecanthropus*. In his lengthy consideration of the larger picture of human descent, in which he invoked the names of virtually every fossil primate then known, including such spectacularly irrelevant forms as the recently extinct Malagasy lemurs *Megaladapis* and *Archaeolemur*, he granted *Pithecanthropus* hardly more than a passing mention. Boule was particularly fond of pointing out that physical resemblance was not necessarily proof of phylogenetic descent. He had used this weapon with great effect to exclude the Neanderthals from the direct human lineage, and he employed it once more

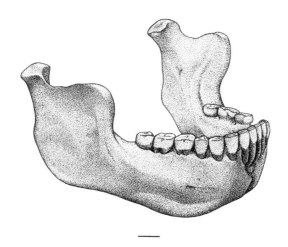

View of the mandible found at Mauer, Germany, in 1908. Type specimen of Homo hei-delbergensis. *Scale is 1 cm.* DM.

to exclude *Pithecanthropus* from any human affinity whatever. According to Boule *Pithecanthropus* was, as Virchow had suggested sixteen years earlier, a giant gibbon. It was necessary to look elsewhere for early humans, and Boule had two suggestions.

The first of these was *Homo heidelbergensis*, a species created in 1908 by Otto Schoetensack for a lower jaw that had been found in a sand quarry at Mauer, near Heidelberg. By this time most scientists had come to accept the sequence of four glacial advances during the Pleistocene proposed in 1894 by the Scottish geologist James Geikie. These cold "glacials" had been separated by three "interglacial" phases of milder climate. On the basis of its associated fauna Schoetensack assigned the Mauer specimen to the Lower Pleistocene, and specifically to the first interglacial period. This made it indisputably earlier than any other human fossil known from Europe. Schoetensack looked upon the specimen as a precursor of the geologically younger Neanderthals, and Boule generally concurred. For while he considered the teeth to be just like those of modern humans, the jawbone itself was in his view apelike. To Boule the ensemble, though inadequate for definitive analysis, was a potential ancestor for *Homo neanderthalensis*. His candidate for the honor of modern human ancestry, in contrast, was *Eoanthropus dawsoni*.

Across the English Channel, interest had lately burgeoned in the possible existence of extremely ancient humans of modern type. In 1910 the influential anatomist Arthur Keith restudied a modern-looking human skeleton that had been found in 1888 at Galley Hill, in deposits of early Pleistocene age. This later turned out to be a much younger intrusive burial, but especially with the discovery of what seemed to be a similar association at Ipswich in 1911, Keith came to accept the apparent combination of modern anatomy with great antiquity. It was but a short step from this to rejection of the Mauer specimen and the Neanderthals as members of the direct line of human descent. For if the evidence from Galley Hill was to be believed, they had coexisted with people of modern kind. By the end of 1912 Keith was strongly committed to this view. And then the Piltdown fossils were announced.

As early as 1908 workers digging in a gravel pit at Piltdown, in Sussex, purportedly handed some fragments of a thick human skull to Charles Dawson, a local lawyer and amateur paleontologist. Over the next year or two various other bits and pieces came into Dawson's hands. In 1912 Dawson gave these specimens to Arthur Smith Woodward, the Keeper of Geology at the British Museum (Natural History) and a leading expert on fossil fish. Smith Woodward joined Dawson and the Jesuit paleontologist and theological mystic Pierre Teilhard de Chardin in working at the site, and together the trio found more bits of cranium, together with part of the right half of an apelike lower jaw, assorted mammal fossils and a few crude stone tools. The fauna seemed to suggest an early Pleistocene or even a late Pliocene age for the site; an estimate that later gained apparent support from geological correlations. Late in 1912 Smith Woodward pieced the various bits and pieces together in a reconstruction of the skull. This was a rather bizarre-looking affair with an ape's jutting jaw and a high human braincase of some 1,070 ml capacity, thus somewhat smaller than typical of modern humans. Just before Christmas he unveiled this marvel, which he named *Eoanthropus dawsoni* ("Dawson's dawn man"), at a meeting of the Geological Society of London. At the same time the neuroanatomist Grafton Elliot Smith analyzed a cast of the inside of the braincase (which more or less reflected the outside shape of the brain). He found it to be quite "simian." This, he said, was only reasonable given that the human brain had evolved from that of an ape; and he concurred with Smith Woodward that here was the ancestor of modern humans.

Arthur Keith was less enthusiastic. For if, as he thought, the Piltdown fossils were no older than his modern-looking Galley Hill and Ipswich specimens, they could hardly represent an ancestor of modern humans. Moreover, he had considerable reservations about Smith Woodward's reconstruction. Within a few months he had come up with his own, which considerably increased the volume of the cranial vault and even gave the apelike lower jaw a human chin. This difference was perhaps hardly surprising, since the Piltdown remains certainly left enormous room for con-

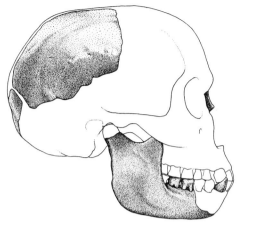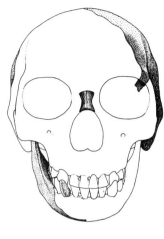

Front and side views of the Piltdown "skull" as reconstructed by Arthur Smith Woodward in 1913. Scales are 1 cm.   DM.

jecture. Most of the forehead and the right side of the skull were missing, and the jaw lacked not only the point at which it had articulated with its cranium, but also the front portion, which would have shown whether or not it had had a chin (modern humans do; apes—and Neanderthals—don't) or a large (apelike) or small (humanlike) canine tooth. Smith Woodward's riposte to Keith's challenge was simple. In 1913 he announced that Teilhard de Chardin had found a lower canine tooth in the gravel pit at Piltdown. And this canine, presumed to be associated with the rest of Piltdown Man, was not only apelike but remarkably similar to the one he had anticipated in his reconstruction. This appeared to settle the matter in his favor and by 1915 the consensus in England was that Piltdown, with a small but humanlike braincase and an apelike jaw, indeed represented the ancestor of modern humans. Keith, however, maintained his belief that its big brain (as he saw it) notwithstanding, Piltdown, like the Neanderthals, represented a dead end in human evolution. It was not until 1917, when Smith Woodward announced that before his death the previous year Dawson had discovered fragments of a second Piltdown individual at a site some distance away but on the same stratigraphic level, that Keith yielded to the majority view.

Or, more properly, he yielded to the English majority view. Despite Boule's early imprimatur, later replaced by the view that the cranium and jaw were unassociated, Piltdown had never found unanimous support on the European continent or in America. As early as 1915 the American mammalogist Gerrit S. Miller had concluded that the Piltdown specimen combined the cranium of a human and the jaw of a chimpanzee. However, a new reconstruction of the Piltdown skull in 1922 by Elliot Smith tended to tip international opinion (including Boule's) in its favor once more. This reconstruction tended to satisfy most parties by producing a compromise brain size of some 1,200 ml, about midway between Smith Woodward's and Keith's estimates. And with it the popularity grew of the "presapiens" theory, which held that at some remote point, probably in the Pliocene, a split had occurred in the human lineage. One branch had given rise to the early appearance of modern humans, via Piltdown. The other branch had led to the doomed Neanderthals. Bolstered by the Piltdown fossil, this scheme appeared so persuasive that by the early 1920s almost no paleoanthropologist—not even Gustav Schwalbe himself, who had succumbed to Boule's influence before his death in 1917—supported the German savant's scheme whereby it was from the Neanderthals that modern humans had stemmed. With the notable exception of one.

From the outset, the Czech-born American physical anthropologist Aleš Hrdlička had viewed the material from Piltdown with the deepest suspicion. Firmly wedded to the view that the human lineage had undergone a steady progressive transformation from an as yet undiscovered "anthropoid precursor" and that modern form had emerged only at the end of the Pleistocene just prior to its spread into the New World, Hrdlička was affronted by Piltdown's combination of apelike and modern human characteristics. Ultimately, while rejecting the claimed antiquity of the cranium, he accepted the jaw as a genuine fossil whose apelike characters were appropriate to its great age, and which in turn provided a link with the still earlier fossil ape *Dryopithecus*. His primary concern, however, was to demonstrate the place of the Neanderthals in the direct line of human descent, and in 1927 he culminated a decade and a half of effort in this direction when on a visit to London he addressed the Royal Anthropological Institute in a lecture entitled "The Neanderthal Phase of Man." Defining the Neanderthals as "the Man of the Mousterian Culture," Hrdlička attacked the notion that these humans had been replaced by invading Aurignacians (the people of the first period of the Upper Paleolithic). Instead, he claimed, the Mousterian had not simply preceded the Upper Paleolithic but had evolved into it. He emphasized the variability in form that existed in the known sample of Neanderthal specimens and suggested that this reflected an ongoing adaptive change in the population—one that continued, indeed, in contemporary human populations, in the form of such things as tooth-size reduction. "There appears to be," Hrdlička concluded, "less justification in the conception of a Neanderthal *species* than there would be in that of a Neanderthal *phase* of man."

But although Hrdlička was echoing an earlier orthodoxy, and his arguments were to survive to be trotted out again almost verbatim in later years, the immediate impact of his broadside against the "presapiens" idea was almost nil, even when he repeated it at book length in 1930. This may have been due to the fact that at this point the Piltdown specimens held center stage. And this in turn suggests the significance of these curious remains: they, more than anything discovered before, established the importance of the fossil record in documenting humanity's place in nature. Of course, as everyone now knows and—it's been claimed—some may have done as early as 1913, the Piltdown fossils were a hoax. The jaw was indeed an ape jaw, and the cranial fragments, though thick, were those of a modern human. The teeth were filed down, and the bones were stained to look like real fossils before being planted in the pit at Piltdown, along with the mammal fossils and the stone tools, for the paleontologists to find. Who the hoaxer was will continue to be debated: almost every possible name has been mentioned, although Dawson was pretty clearly involved in some way or another. The only certain thing is that the perpetrator(s) understood the British paleoanthropological establishment well enough to know what its members would accept relatively uncritically. And although the anomalous Piltdown Man became increasingly peripheralized as other human fossils accumulated, and was widely ignored by the time its fraudulence was proven in 1953, there is no doubt that for a while it did seriously impede progress in the science of paleoanthropology.

Paradoxically, however, it was Piltdown, in its role as "missing link," that for the first time brought the human fossil record squarely into the public eye and established it as a major source of media interest. And it was Piltdown that established the central place of the fossil record in understanding the mysterious process by which modern humans had emerged from an "ape" ancestry. This much is unquestionably on the plus side; but it's also undeniable that paleoanthropology is unique among the subdisciplines of paleontology in ascribing an almost iconic significance to each new fossil that appears. Rarely do paleontologists working on other groups of organisms feel it necessary to undertake wholesale overhauls of their beliefs every time new fossil members of those groups come to light. Paleoanthropologists often do, however, and it's fair to claim that this has been detrimental to their science. But unusual as this paleoanthropological tradition might be, it equally adds excitement and at least the appearance of progress to the study of human evolution—which is maybe what we want most out of the study of our own ancestry. And for better or for worse, this way of doing business can be traced back to the embarrassing Piltdown "specimen."

5

Out of Africa . . .

While Piltdown was riding high, and Dubois was once again permitting *Pithecanthropus* to see the light of day, a young Australian anatomist named Raymond Dart traveled from London to take up the Chair of Anatomy at the University of the Witwatersrand (informally known as "Wits") in Johannesburg, South Africa. Several years before Dart's arrival at Wits in early 1923 a fairly complete fossil cranium of a large-brained extinct human had been found, together with various human and other mammal bones and some stone tools, at the site of Broken Hill (now Kabwe), in Rhodesia (now Zambia). Describing these specimens in late 1921 as belonging to a new species, *Homo rhodesiensis*, Arthur Smith Woodward noted resemblances in the face to the Neanderthals and in the skull roof to *Pithecanthropus* (though the brain was considerably larger). Nonetheless, he found this form to be more like modern humans than the Neanderthals were, with a lighter build to its limb bones. Further, he remarked that in the light of Grafton Elliot Smith's suggestion that "refinement of the face was probably the last step in human evolution," the Broken Hill fossil might "revive the idea that Neanderthal man is truly an ancestor of *Homo sapiens*; for *Homo rhodesiensis* retains an almost Neanderthal face in association with a more modern brain-case and an up-to-date skeleton. He may prove to be the next grade after Neanderthal in the ascending series." Not a word of comparison with Piltdown! Perhaps this was because on the evidence of the associated mammalian bones, which reportedly all belonged to extant species, he believed the specimen to be comparatively recent: filling of the cave "may not even", he wrote, "have been so remote as the Pleistocene period." Be this as it may, apart from a rather enigmatic and highly incomplete cranium found at the South African site of Boskop in 1913, and an apparently ancient

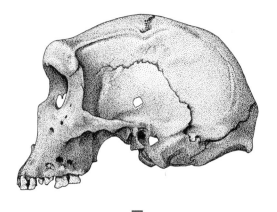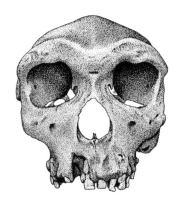

Front and side views of the "Rhodesian Man" cranium discovered in 1921 at Broken Hill (now Kabwe), Zambia. Scales are 1 cm. DM.

but modern-looking skeleton recovered at Olduvai Gorge in Tanganyika (now Tanzania) in the same year, Broken Hill represented the sum total of human fossils known at that time from Africa. As far as human paleontology was concerned, Europe and Asia still monopolized the limelight.

During his time as a junior faculty member at University College London, Dart had worked with Elliot Smith and had absorbed some of his mentor's fascination with the evolution of the human brain. But nothing was further from his mind than the discovery of human fossils when, in 1924, one of his students, Josephine Salmons, brought him a fossilized baboon skull. This had been blasted out of the Buxton lime quarry near a place called Taung, some considerable distance to the northwest of Johannesburg. His colleague R. B. Young, professor of geology at Wits, volunteered to bring him more specimens from this site, and into Dart's hands there duly came a fossil consisting of the face and an endocranial cast—a natural mold of the inside of the skull—of an immature hominoid, clearly either an ape or an apelike human. As a neuroanatomist, Dart found his attention first attracted to the brain cast—which in any case, unlike the face, was not enclosed in and obscured by a hard limey matrix. True, Dart's former boss Elliot Smith had studied plaster endocranial casts of various recent and fossil humans and other mammals and had received a less than ecstatic response. For example, it was shown that the convolutions of the brain were often poorly reflected in casts of the inside of the skull. Moreover, most of his colleagues held that inferences of the kind that Smith was willing to make about mental and motor capacities would be hard to justify even where the brain itself was available for external examination. But undeterred by Smith's experience, Dart allowed his appraisal of the brain cast, bolstered by other features of the specimen, to lead him to a far-reaching conclusion. This he arrived at without the usual panoply of scholarship: library facilities were inadequate, he lacked appropriate comparative specimens, and he worked on his specimen alone, in isolation from colleagues.

Once he had laboriously removed the tough matrix from the face of the "Taung child" (though he had not as yet freed the lower jaw from the upper), Dart immediately prepared a report which he sent to the London journal *Nature*. This appeared on January 6, 1925, and its title, "*Australopithecus africanus*: The Man-Ape of South Africa," made its message clear. Dart pointed out several features of the face that were obviously humanlike, among them the high, rounded forehead unadorned by eyebrow ridges, the subcircular orbits, the delicate structure of the cheekbones, the flattish profile, and the lightly built lower jaw. He also considered the teeth (which corresponded to those of a modern six-year-old, with just the first permanent molar in place) to be "humanoid." The milk canines barely projected beyond the other teeth, the incisors were small, there was no diastema (a gap between the canine and front premolar of the ape lower jaw which receives the large upper canine), and so forth. Most of these characters were actually related more than anything else to the tender age of the individual, for in both apes and humans the juvenile face is small relative to the braincase, and delicately built. The brain cast preserved the entire right side of the brain, with minor damage here and there. It was small (modern estimates put the volume of the whole thing at 440 ml), and Dart confessed that in the adult it would probably not have surpassed a gorilla's in size. Nonetheless, he saw in it evidence for a greater expansion than in the apes of the higher centers of the brain, with a "rounded and well-filled-out contour, which points to a symmetrical and balanced development of the faculties of associative memory and intelligent activity." Most significant, he thought, was the humanlike backward displacement of the lunate sulcus, a groove which marks the front of the brain's primary visual cortex. What such displacement would reflect is an increase in the extent of the "association" cortex that lies in front of the sulcus, and this suggested to Dart that his creature had "laid down the foundations of that discriminative knowledge of the appearance, feeling and sounds of things that was a necessary milestone in the acquisition of articulate speech." It had also, he believed, acquired the habit of upright walking, since the inferred position of the foramen magnum, through which the spinal cord exits the brain, was right underneath the skull, rather than toward the rear as in the quadrupedal apes.

In 1925 ways of assessing relationships between organisms were rather

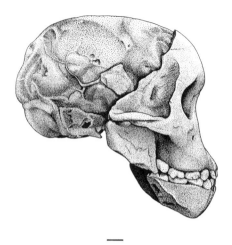 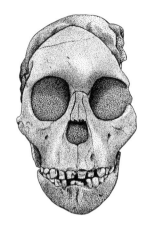

—   —

Front and side views of the face and brain cast of the "Taung Child," found at South Africa's Buxton limeworks in 1924. Scales are 1 cm.   DM.

rudimentary, based primarily on overall resemblance—which generally corresponded to the intuition of the observer. It was not yet realized (and was not to be for many decades) that terms such as "apelike" and "humanlike" were by themselves essentially meaningless when it came to assessing taxonomic relationships. Dart cannot be faulted for analyzing his fossil in these terms; but reading between the lines of his description suggests that his basic attitude toward the specimen was founded in a tradition that was archaic even in his day. "The whole cranium," he wrote, "manifests in a striking degree the *harmonious relation* of calvaria [braincase] to face emphasised by Pruner-Bey." This reaction harks back to the essentialist aesthetic that had informed reaction to the earliest human fossils to come to light. This had emphasized the "brutishness" of the apes and even of the "lower races" of humans, in contrast to the civilized elegance of the "higher races"; and indeed the French savant Pruner Bey quoted by Dart has gone down in history as the man who wrote the Neanderthaler off as a strongly built Celt, "somewhat resembling the skull of a modern Irishman with low mental organization." It may be of significance in this regard that Dart himself was no fan of *Pithecanthropus*, which he described as "a caricature of precocious hominid failure."

But although it was the "humanoid" aspects of his specimen that most impressed Dart, the name he chose for its species, *Australopithecus africanus*, translates as "southern ape of Africa." Indeed, he described it himself as a "man-like ape," a representative of "an extinct race of apes *intermediate between living anthropoids* [apes] *and man.*" This "race," while "well advanced beyond modern anthropoids in just those characters, facial and cerebral, which are to be anticipated in an extinct link between man and his simian ancestor," had at the same time a small sized brain "lacking the distinctive, local expansions which appear to be concomitant with and necessary to articulate man." It was thus "no true man," and Dart proposed a new family, Homo-simiadae, to contain it. Noting that the site of Taung was right on the edge of the Kalahari Desert, and believing that the harsh climate of the area had been stable back into the remote past, Dart proposed that "for the production of man a different apprenticeship [from that served by the apes in the 'luxuriant forests of the tropical belts'] was needed to sharpen the wits and quicken the higher manifestations of intellect—a more open veldt country where competition was keener between swiftness and stealth, and where adroitness of thinking and movement played a preponderating role in the preservation of the species." I love this last passage, though Dart was to be roundly condemned by his colleagues for excessive rhetoric. For, as he reveals himself here, Dart was the first student of the human fossil record who seems to have felt viscerally the drama of the biological history of mankind. He was certainly among the first to perceive that the story of our species is one of individuals, populations, and species living, striving, and dying as part of an enormously complex web of organic life, and to see that this story is not complete if we do not let ourselves go well beyond the teeth and bones that are our primary evidence of it. Dart was to allow his vivid imagination to get the better of him on later occasions, but here I cannot help but feel that even though he was writing in the august pages of *Nature*, he hit a needed note.

Few at the time agreed. In England, either you are "sound" or you are not. And according to the distinguished anatomist Wilfrid Le Gros Clark, Dart's soundness had been in question ever since, before leaving University College, he had coauthored a paper which gave the impression that he "might be inclined too hastily to arrive at conclusions on too little evidence." And many felt, on reading his *Nature* article, that his purplish prose might hint once more at overenthusiastic conclusions. The week following the publication of his article, four of the greats of British paleoanthropology delivered their judgments in *Nature*. These were based, of course, entirely on Dart's description and illustrations, for none of these scholars had seen the original specimen or even a cast (replica) of it. Casts, in fact, were slow to follow (nobody in Dart's department knew how to make them, and eventually a professional plasterer was hired for the job!), and first reached England not for the delectation of Dart's colleagues, but for display to the public at the British Empire Exhibition that opened in the summer of 1925.

The British paleoanthropological establishment, forced to peer at the specimen through glass while being jostled by the passing hoi polloi, was not amused.

But given the circumstances, the generally unfavorable reaction among Dart's colleagues was understandable. It is always hard to know what to make of a fossil when only photographs and someone else's description are available; many of the "humanoid" features to which Dart had pointed might as well have been attributed to the tender age of his specimen; the features of the brain cast are still being argued over to this day; and Dart had given not one hint as to the geological age of his fossil—though it was clearly old, whatever that might mean. So most initial reactions were perhaps justifiably cautious. Arthur Keith inclined toward placing *Australopithecus* in the same subfamily as the African apes. Elliot Smith was intrigued but reserved his opinion, particularly pending the opportunity to evaluate the teeth once the lower jaw was detached from the upper face. Smith Woodward, champion as ever of Piltdown, dismissed Dart's specimen out of hand, claiming that it "certainly has little bearing on the question" of whether humans evolved in Asia or Africa. Finally, the Cambridge anatomist W. L. H. Duckworth, on whose treatise *Morphology and Anthropology* Dart had relied heavily in comparing the Taung infant to the apes, admitted certain advanced features of the skull but found its closest resemblance to be with the gorilla. Overall this response was pretty tepid; but it was certainly preferable to the public reaction to newspaper coverage of Dart's claims. One correspondent accused Dart of being a Priest of Baal; someone wrote from France to inform him that he would "roast in the quenchless fires of Hell"; an Englishman hoped to see him in "an institution for the feebleminded."

For some reason (probably because at heart he was interested in brains, not bones; he had, after all, only stumbled into paleoanthropology by accident) Dart did not seek at the time to bolster his case by looking for adult *Australopithecus* fossils at Taung or at any of the other lime mines that dotted the landscape of central southern Africa. And although his university almost immediately offered him the opportunity to travel to show his prize to his colleagues and to compare it directly with specimens in the great natural history institutions of Europe, it was not until 1931, by which time attitudes had generally hardened, that he brought the Taung juvenile to England, where it could be examined at firsthand by his colleagues. By then, moreover, paleoanthropological attention had been attracted away from Africa, once more toward Asia.

For some time paleontologists had begun to suspect that climatic change might have been the factor that initiated the divergence of the human lineage from that of the apes. And in certain influential quarters it was believed that the most likely locus of such change was central Asia, where uplift of the Himalayas had combined with a general post-Eocene trend toward climatic cooling and drying to create open high plains. Any ape living in this part of the world would have had to abandon a forest-living existence in favor of one in a more open-country habitat, perhaps with the consequences hinted at by Dart in his discussion of *Australopithecus*. The principal proponent of this viewpoint was Henry Fairfield Osborn, president of the American Museum of Natural History. This distinguished paleontologist was so convinced that central Asia would prove to be the cradle of mankind that during the 1920s the American Museum sponsored a series of remarkable expeditions to the region. There, Osborn believed, "while the anthropoid apes were luxuriating in the forested lowlands of Asia and Europe, the Dawn Men [had been evolving in the invigorating] atmosphere of the relatively dry uplands." In the event, while the expeditions made many extraordinary discoveries they failed to find any human fossils; but they did have the effect of drawing public and paleoanthropological attention towards this part of the world. The fact that Osborn was at the same time tending more and more to reject an ape ancestry for humans, or at least to push it as far back into the remote past as possible, detracted not at all from the attractiveness of the central Asian scenario.

As it happened, however, by the time the American Museum expeditions got under way paleontological work had already been started in China, principally as a result of the zeal of Johan Gunnar Andersson, a mining engineer by trade but a paleontologist by avocation. Through his efforts, a monopoly on fossil-seeking in that country had already been obtained by scientists affiliated with the University of Uppsala, Sweden. Thus excluded by official agreement from working in China, the Americans confined their attention to the uplands of Mongolia, where the rocks were so ancient that nobody without Osborn's peculiar viewpoint on the matter would have expected to find the remains of early humans—though, as we've seen, much else of great interest was found. The Europeans, on the other hand, directed their attention among other things toward more recent cave deposits, from which "dragon bones," otherwise fossil mammal teeth, had long been known. In this they received valuable advice from the distinguished paleontologist Walter Granger, an advance man for the American Museum team. One of the sites they explored, an abandoned lime mine not far from Beijing, was Chou K'ou Tien (now Zhoukoudian).

This work was not long in yielding results. The fill from the Zhoukoudian cave was rich in fossils, and by the summer of 1921 it had already yielded the first direct evidence (Andersson had already identified correctly the rather crude stone tools that were abundant at the site) of what was to become known as "Peking Man." But for his own reasons its discoverer, a young Austrian paleontologist named Otto Zdansky, kept the discovery secret until 1926. In that year, at a meeting held in Beijing, he described two human teeth from the fossil assemblage at Zhoukoudian. Zdansky himself declined to make much of this find, but Davidson Black, a Canadian who was then professor of anatomy at the Peiping Union Medical College, was fired with enthusiasm. He arranged financing from the Rockefeller Foundation, sponsors of the Medical College, and new excavations began at Zhoukoudian in the spring of 1927, under the field direction of a young Swede named Birger Bohlin. By the fall of that year, another human tooth had been recovered. Black had regarded the two original teeth as belonging to the genus *Homo*, but on the basis of this new discovery he created the new genus and species *Sinanthropus pekinensis*, the "Chinese man of Peking." Few were impressed. One tooth, which was all Black had to show his colleagues when he visited Europe (the original two teeth were in Sweden), was not much on which to base an entire new genus of humans. But Black persevered, and with continuing Rockefeller support he established the Cenozoic Research Laboratory in Beijing, whose first project would be to enlarge the excavations at Zhoukoudian, using a largely Chinese team.

Enormous quantities of the cave deposit were removed and processed, and thousands of mammalian fossils were recovered, among them a few more human teeth. But it was not until the end of 1929 that these labors paid off with the recovery of a braincase of *Sinanthropus*. This was the beginning of a stream of new discoveries that continued beyond Black's early death in 1934, until guerilla activity around Zhoukoudian brought work there to a halt in 1937. By that point, the site had yielded fourteen partial skulls as well as various bones of the postcranial skeleton of *Sinanthropus*. All of this engaged the interest of an eager public—whose attention was thereby distracted from *Australopithecus*—but from the moment that the Chinese paleontologist W. C. Pei had recovered the first braincase in 1929 it was clear that here was an early human very much like the one from Java. The forehead of the first Zhoukoudian braincase—like later ones—was a little steeper than in Dubois's *Pithecanthropus* from Trinil, and the brain capacity a little larger; but Black was more impressed by the similarities than the differences between this pair of early humans, though he continued to use the name *Sinanthropus*, implying that the two fossils indeed belonged to different genera. For Black, *Sinanthropus* was a more advanced form, occupying a position intermediate between *Pithecanthropus* and the Neanderthals.

Predictably, Eugene Dubois disagreed with all this, if in an odd way: *Sinanthropus*, he thought, was "perfectly" human—probably, indeed, a Neanderthal, though he later retracted this idea—while *Pithecanthropus* showed some apelike characters, especially in the brain. To maintain the

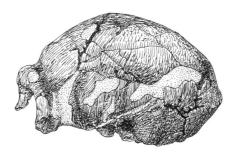

—

Side view of the most complete of the "Peking Man" crania (Skull XII) discovered in the prewar period at Zhoukoudian, China. Scale is 1 cm.   DS.

integrity of his theory of brain evolution, Dubois was concerned above all to show that *Pithecanthropus* did not occupy the same level of evolutionary development as either humans or living apes, and the advanced features to which the describers of *Sinanthropus* pointed thus forced him to distance the latter from his own fossil. To the world, however, *Sinanthropus* convincingly demonstrated that *Pithecanthropus* was indeed an early human, and as material from Zhoukoudian accumulated, most authorities soon came to believe that both represented the same group of early humans. Particularly influential in this regard was the fact that by the end of 1934 a series of skullcaps had been found at Zhoukoudian in which brain size varied from 850 to 1,200 ml, a range within which the Trinil skullcap comfortably fitted.

One thing that set off the Zhoukoudian finds from those in Java was a fuller archaeological context. In 1931, Black reported that some of the animal bones from Zhoukoudian had apparently been charred by fire, and that tests had shown that some blackened layers of the cave deposits contained quantities of carbon. No actual hearths were found, as had been in some of the later European cave sites; but his evidence was enough to convince Black that *Sinanthropus* had included the control and use of fire among its behaviors. In the same year W. C. Pei had reported finding crude quartz and other stone artifacts at Zhoukoudian, so Peking man was rapidly acquiring many of the behavioral traits generally associated with humans. By 1939, moreover, Black's successor, the German anatomist Franz Weidenreich, had added the less attractive traits of murder and cannibalism to the relatively anodyne pursuits of tool and fire use. Weidenreich noted that the remains of almost forty human individuals, fifteen of them children, had been found in the cave, but that there was not one complete skeleton. Indeed, the fossil human remains were overwhelmingly cranial, and all were fragmentary, many bearing apparent evidence of physical trauma while still covered in soft tissues: witness to their bearers' having "suffered violent attacks". All of the Zhoukoudian bones, human and non-human alike, were thought by Weidenreich to be the remains of *Sinanthropus*'s meals. Further, the bases of all the human crania were broken, presumably for cannibalistic removal of the brain within. This is an explanation evoked on numerous occasions before and since to explain damage to the bases of fossil human skulls, and Weidenreich was particularly impressed by the claims of the anthropologist Paul Wernert that, from prehistoric times to the present, "there had always existed an entanglement between the two rites of anthropophagy and head-hunting," based on the notion of "increasing the material and spiritual qualities of an individual or the community taking possession of the corpse of the vanquished." Taking all these indications together, Weidenreich could not avoid the conclusion that "the *Sinanthropus* population of Choukoutien had been slain and that subsequently their heads were severed from the trunk, the brains removed and the limbs dissected."

While realizing that "it may prove rather distressing to some sensible people to hear that the most primitive ancestor of recent mankind had been responsible for such terrible acts as manslaughter of women and children and cannibalism", Weidenreich regretted that his audience would not find the behavior of other fossil humans "any more pleasing." "Ten years ago", he wrote,

> I was already able to show that the fossil man of Weimar-Ehringsdorf, a representative of the Neanderthal group of the last interglacial period, must have indulged in similar customs. The skull was broken like those of *Sinanthropus*, the frontal bone shows very characteristic markings of heavy blows, and the base is missing . . . The remains of the Krapina population, who lived during the same period as the man of Weimar-Ehringsdorf, were so completely broken that it was even impossible to reconstruct one entire brain case from the numerous bone fragments belonging to some 20 individuals. . . . The recently discovered skull of Steinheim, who may be even older than the Weimar-Ehringsdorf and Krapina man, also shows . . . manipulations similar to those shown in *Sinanthropus* skulls.

Some disagreed, as would many today, if not always for the same reasons. Marcellin Boule, for example, felt that *Sinanthropus* was entirely too primitive to have made the stone tools, or to have lit the fires, or to have hunted the animals, of Zhoukoudian. In his view another and more ad-

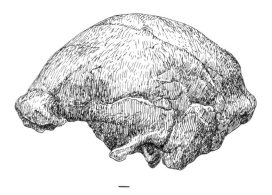

—

Side view of one of several crania (Skull V) discovered at Ngandong (Solo), Java, during the 1930s. Scale is 1 cm.   DS.

vanced human must have lived there and have been responsible for the ashes, the tools, and the accumulation within the cave of the bones of *Sinanthropus* and other animals. He did not find it surprising that there was no trace of this advanced human in the fossil record of Zhoukoudian, for, as he pointed out, there were plenty of other fossil sites with evidence for human occupation, but no human bones. Nonetheless, Weidenreich's interpretation generally prevailed—and, as it happened, foreshadowed a more comprehensive scenario of a bloodstained human past that was shortly to emerge from Africa.

As work continued in China, attention was once again being paid to the fossil potential of Java. In 1931 and 1932 the Dutch mining engineer W. F. F. Oppenoorth recovered a series of eleven crania, varying in brain capacity from 1,035 to 1,255 ml, from a site in the Solo Valley of western Java. These fossils were reckoned to be of late Pleistocene age, thus younger than the early Pleistocene date the Zhoukoudian fauna seemed to suggest for the Peking Man specimens; but they resembled the latter in certain ways, although Oppenoorth was initially more impressed by their similarities to the Neanderthals. And while Oppenoorth later preferred to emphasize the differences between the Solo skulls and those of the Neanderthals, many came to see *Javanthropus* (or *Homo*) *soloensis* as a sort of Asian Neanderthal equivalent.

Discoveries in Java that received considerably more attention were made by the German paleoanthropologist Ralph von Koenigswald in 1936 and the years following. The first human fossil that von Koenigswald recovered was the cranium of a child, which he believed to be of early Pleistocene

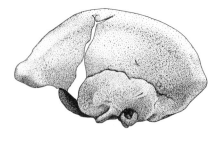
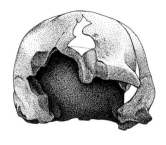

Front and side views of the braincase ("Pithecanthropus II") discovered at Sangiran, Java, in 1937. Scales are 1 cm.   DM.

age, at a place called Modjokerto. He clearly thought this young specimen to be a *Pithecanthropus*; but in deference to the opposition of Dubois, who likened it to the Solo skulls which he believed to be ancestral to the aborigines of Australia, von Koenigswald named his new find *Homo modjokertensis*. He did not, however, extend the nomenclatural courtesy to a very robust jaw fragment he found close to the village of Sangiran. This he squarely placed in *Pithecanthropus*, with which he considered it coeval. By 1937 von Koenigswald had discovered a skullcap in nearby sediments at Sangiran. The specimen was more complete than Dubois's Trinil fossil, but in comparable features the two were pretty alike, proving beyond doubt that *Pithecanthropus*, like *Sinanthropus*, lay in the human family. Von Koenigswald estimated its cranial capacity at about 750 ml; more recent estimates make it a little larger than this but still smaller than the Trinil specimen. Partly because of the smallish brain size, von Koenigswald considered his *Pithecanthropus* to be more primitive than *Sinanthropus*. These initial finds were followed over the next several years by others, including the back part of another skullcap with an associated maxilla, and a massive mandible. Various names were assigned to this material, but differences in size and robusticity between different specimens were eventually ascribed to sexual dimorphism—size and shape differences between males and females.

Unfortunately, geological controls on the excavation of all of the Java fossil collections were extremely poor. Hence the dating of the Trinil, Sangiran and Solo specimens remains debated today, though very recent developments indicate that an extraordinarily long time may have elapsed between the earliest materials from Trinil and the latest from Solo.

In 1939 von Koenigswald journeyed to China to compare his fossils with

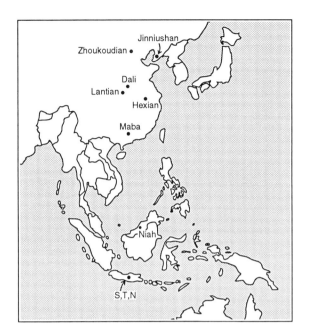

Map showing the locations of the major sites of human fossil discoveries in eastern and southeastern Asia. S,T,N denotes the approximate location of Sangiran, Trinil, and Ngandong. DS.

the *Sinanthropus* materials, by then under the care of Franz Weidenreich. Following the cessation of excavations at Zhoukoudian in 1937, Weidenreich had devoted himself to intensive anatomical study of the *Sinanthropus* specimens. He had already begun to form an explicit theory of human descent, in which the various human races could be traced far back in the fossil record. Weidenreich had done this partly to provide an alternative to the old polygenist view of human origins, which he dismissed out of hand although he saw plenty of evidence in the fossil record for a "polycentric" origin of modern mankind. Each race, he thought, had evolved separately toward *Homo sapiens*, following some form of inner directionality linked to enlargement of the brain. And even at a time when developments in evolutionary theory were making such ideas distinctly outmoded, Weidenreich was not unwilling to explain this directionality in terms of orthogenesis, the notion that evolution is somehow goal-oriented. In this broader scheme Weidenreich placed *Sinanthropus* as a precursor specifically of the "Mongol" (Asian) race.

To Dubois's dismay, before von Koenigswald's arrival in Beijing Weidenreich had already found points of distinction between the partial femora from Zhoukoudian and the femur from Trinil. These had persuaded him that the Trinil skullcap and femur were not associated with each other, though he accepted a general affinity between the species represented by the skullcap and that from Zhoukoudian. On the other hand, he was initially impressed by the resemblances between the Sangiran material and the Solo skulls. He suggested that at Sangiran von Koenigswald might have found female members of the species represented at Solo, and he contested the view that *Pithecanthropus* was more primitive than *Sinanthropus*. Direct comparison of the Chinese and Javanese specimens, however, persuaded both Weidenreich and von Koenigswald beyond doubt that these "prehominids" were "related to each other in the same way as two different races of present mankind, which may also display certain variations in the degree of their advancement." As to which was the more "advanced" they refused to be categorical, though they did agree that "Pithecanthropus shows some significant characteristics which must be considered more primitive than those evident in Sinanthropus."

Interestingly, while they continued to use the terms "Sinanthropus" and "Pithecanthropus" in their joint report to *Nature*, von Koenigswald and Weidenreich did not italicize them. This is obligatory for formal zoological names, and both authors had adhered to this convention when referring to the respective forms in earlier publications. Moreover, in their joint paper they italicized *Homo soloensis*, which they considered to be the next step in human evolution. This suggests that they were using these genus names (as I shall from now on) simply for the sake of convenience, and that if pressed they would have allocated Java and Peking Man to *Homo* as well. Certainly Weidenreich went on to develop his theory of a Sinanthropus-Mongol connection to state specifically that "at the very appearance of true hominids [represented by Pithecanthropus and Sinanthropus there must have existed several] branches, morphologically well distinguishable from one another, which all proceeded in the same general direction with mankind today as their goal." Pithecanthropus, he believed, was on the line, via *Homo soloensis* and "Cohuna Man" (a relatively recent fossil Australian), to modern aboriginal Australians; modern Chinese are derived from Sinanthropus via intermediates as yet unknown; later African ancestry was represented by the "Neanderthaloid" Rhodesian Man; and while Neanderthals appeared to have been replaced in western Europe by invading moderns, it was not out of the question that they had evolved elsewhere into the form which subsequently supplanted them. In its essentials this scheme has shown remarkable tenacity in surviving right up to the present day, albeit in a somewhat modified form.

Sadly, the Peking Man fossils did not long see the light of day. Davidson

Black had established the Cenozoic Research Laboratory on the understanding that the Zhoukoudian fossils would remain in China, but fears grew for the safety of the specimens as the Japanese encroached on China throughout the period leading up to the outbreak of war in the Pacific. These fears were not unfounded; despite precautions taken by von Koenigswald before his internment following the Japanese takeover of Java, one of the Solo skulls wound up in the Japanese Imperial Palace in Tokyo. Eventually, the possibility was broached of sending the human fossils from Zhoukoudian to a place of temporary refuge in the United States. During most of 1941 the custodians of the fossils dithered over how and when this should be done, and Weidenreich did not take them with him when he left at midyear to take up residence at the American Museum of Natural History. In the event, the crucial decision to export them was not taken until just before the outbreak of formal hostilities between Japan and the United States in December 1941. When, on December 8, the Japanese searched the fossils' repository they found only casts; but what had happened to the original fossils remains a mystery. They appear to have been packed in a couple of footlockers and entrusted to a platoon of U.S. Marines for transportation to the port of Tianjin. There they were to be loaded on the S.S. *President Harrison*, bound for the United States; but the ship was sunk en route to the port, and the fossils simply vanished. To this day the fate of the Zhoukoudian bones remains unknown, though theories abound. Fortunately, though, before his death in 1948 Weidenreich was able to complete a series of exquisitely detailed monographs on Sinanthropus, and what for their period are technically excellent casts still remain to represent the originals, pale substitute though they are. Postwar excavations at Zhoukoudian have produced a few more human fossils (including part of one of the braincases described by Weidenreich), but nothing like the prewar riches.

But whatever the fate of the Peking Man fossils, together with von Koenigswald's specimens they had served the critical purpose of validating Dubois's original finds. It is hard for us to appreciate it today, but as late as 1938 an anonymous commentator was able to write in *Nature* that "of recent years, opinion has tended in an increasing degree to incline" to the view that Dubois's Pithecanthropus remains were ape rather than human. By the time that Sinanthropus disappeared, that had irreversibly changed. No question, then, that the finds from eastern Asia deserved the limelight in which they basked during the 1930s; but in Africa, with less fanfare, equally significant developments were meanwhile afoot.

6

... Always Something New

One of the most remarkable characters in the history of paleoanthropology was Robert Broom. A Scot by birth, a physician by training, and consumed by an interest in the origin of the mammals, Broom traveled in 1892 to Australia, home of the world's most primitive mammals. After five years he moved on to South Africa, which was to be his home for the remainder of his long life. After describing a fauna of small mammals from a Pleistocene bone breccia, Broom shifted his interest to the fossil-rich sediments of the Karroo region of central South Africa. He rapidly became the leading world authority on the mammal-like reptiles that these rocks produced in extraordinary abundance. An early (and for many years the only) whole-hearted supporter of Dart's interpretation of *Australopithecus* as an ape-human intermediate, Broom gave up the practice of medicine in 1934, at the age of sixty-eight, and took up a post at Pretoria's Transvaal Museum. He busied himself with reptilian fossils for his first couple of years at the museum; but in 1936, by which time no effort had been made in twelve years to recover adult *Australopithecus* specimens, Broom resolved to rectify this situation. Short of funds, he looked for cave deposits closer to home than Taung, not a difficult undertaking since caves were hardly lacking in the dolomitic limestones of the Transvaal. Most of these were filled with accumulated rubble cemented together by redeposited lime, and many were the site of limeworks. Broom's attention was rapidly drawn (by Trevor R. Jones, a student of Dart's) to a lime mining operation at Sterkfontein, only about forty miles from his office in Pretoria. Coincidentally, the manager there was the same man who had been in charge at Taung when Dart's juvenile *Australopithecus* had turned up. In August 1936 the site yielded a rather battered and incomplete adult *Australopithecus* skull, which Broom

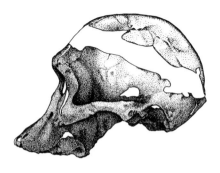
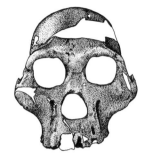

Side and front views of the cranium Sts 5 ("Mrs. Ples"), from Sterkfontein Member 4, South Africa. This is the best-known cranium attributed to Australopithecus africanus. Scales are 1 cm. DM.

immediately reported to *Nature* as confirming Dart's views. The associated fauna suggested to Broom that the Sterkfontein deposit was somewhat younger than the one at Taung, so he placed his find in a new species, *Australopithecus transvaalensis*. Later, impressed by differences that he perceived between the teeth of the Taung and Sterkfontein specimens (though his fossil had only four teeth and the Taung child had erupted only the first permanent molars), he shifted his specimen into a new genus, as *Plesianthropus* ("near man") *transvaalensis*.

There was no doubt in Broom's mind that he had before him the confirmation of Dart's claims that he had been seeking. Among his European colleagues, however, news of his finds was received politely but without enthusiasm. Arthur Keith's dismissal of *Australopithecus* in his *New Discoveries Relating to the Antiquity of Man* of 1930 (which had preempted a long monograph prepared by Dart) was regarded in Britain as the last word on this creature; and the conclusion by the German Wolfgang Abel that the Taung baby was but a juvenile ape was similarly influential on the Continent. In this intellectual climate Broom's fragmentary fossil was regarded as hardly sufficient evidence to justify abandoning the conventional wisdom that *Australopithecus* was no more than an early ape. Broom persevered, however, and over the several months following his initial discovery recovered a few more bits and pieces from Sterkfontein. But his next real break came in June 1938, when a local schoolboy, Gert Terblanche, found a palate with a partial dentition at a neighboring site called Kromdraai, just across the valley from Sterkfontein. Other finds there followed, and pretty soon Broom had assembled half a skull of a form more heavily built than his *Plesianthropus*, with a flatter face and much larger chewing teeth. These differences led Broom to create yet another genus and species, *Paranthropus* ("next to man") *robustus*, and as he worked on the block of matrix from which the first specimens had been removed he found a few bones of the postcranial (body) skeleton, including a very humanlike talus (ankle bone).

In America Broom's views on the protohuman status of *Australopithecus* rapidly gathered support from the influential paleontologist and comparative anatomist William K. Gregory, of the American Museum of Natural History. With his colleague Milo Hellman, Gregory traveled to South Africa in June 1938 to view the original specimens from Taung and Sterkfontein, just as the first fossils from Kromdraai were coming to light. Gregory and Hellman's appraisal of the dental evidence, in particular, led them to the conclusion that here were "in both a structural and a genetic sense the conservative cousins of the contemporary human branch". "The whole world," Gregory told a meeting of the Associated Scientific and Technical Societies of South Africa on July 20, 1938, "is indebted to these two men [Broom and Dart] for their discoveries, which have reached the climax of more than a century of research on that great problem, the origin and the physical structure of man." To Gregory and Hellman the South African fossils were true intermediates between extinct Miocene apes such and *Dryopithecus* and modern humans. In 1939, while reserving judgment on whether *Australopithecus*, *Plesianthropus* and *Paranthropus* actually deserved to be recognized as separate genera, the American scientists formally placed them all in the same subfamily, Australopithecinae*, of the family Hominidae. The other hominid subfamily, Homininae, contained modern humans and their post-australopithecine precursors. The apes, in contrast, occupied their own family, Pongidae (for *Pongo*, the orangutan).

Another early supporter of the human affinities of the australopithecines was von Koenigswald, who declared in 1942 that he could find in the dentition "no characteristic to force this group from the Hominidae." Nonetheless, he could not consider them ancestral to man, partly because be believed they were geologically too young, but also because their teeth were too large. Recognizing that this latter factor was no obvious impediment to ancestral status, von Koenigswald felt obliged to elaborate. First, he said, it was a general law that organisms became larger over their evolutionary histories, as exemplified by the increase in size over time of the human brain. But Pithecanthropus, Sinanthropus, the Neanderthals, and modern

---

*In the jargon of zoological nomenclature, the names of subfamilies always end in "-inae", those of families in "-idae", and those of superfamilies in "-oidea". The name Australopithecinae thus formally indicates a subfamily rank, although the term "australopithecine" is sometimes used informally, as we'll see.

people formed a series in which tooth size decreased, and this required special explanation. This explanation lay in the use of tools to prepare food and of fire to cook it—and for such activities there appeared to abundant evidence from Zhoukoudian on. With these innovations, mankind "used his teeth in a different way from his anthropomorphic ancestors. When he began to talk he also used his jaw muscles in a different way. Only in the development of civilization can we find a reason for the typically human evolution: the reduction of the dentition combined with an astonishing, progressive development of the brain—the two surely interdependent." So far so good—perhaps. But von Koenigswald went on to suggest that because tooth size increase was inevitable in the absence of civilizing factors, Pithecanthropus, probably the first human to bear civilization in his sense, must have had the largest teeth ever possessed by a member of the human lineage. From this it followed that "every human-like being having bigger teeth than Pithecanthropus must be excluded from the direct line of human evolution." The australopithecines, regrettably, fell into this latter category. By the middle of the twentieth century reasoning of this kind would have attracted incredulity in any other branch of paleontology, but it is a testament to the continuing insularity of paleoanthropology that to von Koenigswald's colleagues it did not sound unduly out of place.

Doubtless fortified by the support of Gregory and Hellman—though in any event it was not in his character to be assailed by self-doubts—Broom labored throughout the years of World War II on a monographic treatment of his finds. In this work, which appeared in early 1946 when he was almost eighty, Broom concluded that his australopithecines had walked bipedally and reiterated his and Gregory's conclusion that although they had relatively large faces their dentitions were of human type (teeth are particularly important to paleontologists because, as the hardest tissues of the body, they are more frequently preserved as fossils than other structures). They had small brains, "probably between 460 cc [1 cc-1 ml] and 650 cc" (the former turns out to be about an average figure, the latter something of an overestimate), but they probably had advanced manipulative abilities that embraced tool use. Altogether, wrote Broom of the australopithecines, "if one could be found alive today I think it probable that most scientists would regard him as a primitive form of man". He noted the geological evidence that suggested these creatures had lived in open country; and as to geological age, on faunal grounds Sterkfontein appeared to date from the middle or late Pliocene, Taung possibly from a bit earlier, and Kromdraai from somewhat later, possibly the early Pleistocene. Later humans, Broom believed, were derived from a Pliocene australopithecine not very different from *Australopithecus*.

Impressive though Broom's monograph was, however, it was received by his British colleagues with a notable lack of enthusiasm. According to Wilfrid Le Gros Clark this was partly because it bore certain marks of haste in preparation but principally because it was published along with what was considered to be an overambitious appraisal of the available brain casts, natural and artificial, from all three sites. This had been prepared by G. W. H. Schepers, a colleague and former student of Dart's who claimed to be able to deduce a whole range of humanlike behaviors from the bumps and grooves on the casts. What's more, Schepers vigorously attacked the idea that humans could be descended from "specialized and degenerate anthropoids"; and under the influence of the anatomist Frederic Wood Jones (who had developed his own theory of human descent not from the apes, which he considered an altogether separate lineage, but from an Eocene form resembling the tarsier, a small and enigmatic primate), he argued that the human lineage had been distinct since the Eocene. Schepers' arguments did not go down well, and Broom's simpler if not entirely restrained reasoning was tainted by association. Indeed, it was not until 1947, when Le Gros Clark himself visited South Africa and examined the full range of australopithecine material, that the tide began to turn. Le Gros Clark was the first member of the British anthropological establishment to have studied all these fossils at firsthand, and skeptic though he initially was, he rapidly became convinced that Broom was right. When he took his new message back to England he ran into some opposition from Wood Jones and his fellow anatomist Solly Zuckerman, an instinctive and intransigent reactionary; but his authority rapidly carried the day.

Even Broom's exact contemporary Arthur Keith recanted, writing in 1948 in his *A New Theory of Human Evolution* that "of all the fossil forms known to us, the australopithecinae are the nearest akin to man and the most likely to stand in the direct line of man's ascent." But although he accepted that these forms showed that a large brain must have been acquired in the course of the evolution of the human lineage, rather than having been present at the outset, Keith found himself unable to regard them as early humans. This was because to him, as to many others, small brains were by definition apelike, and the presence of small brains in early members of the human lineage made it necessary to define a "cerebral Rubicon" in brain size: a threshold which had to be exceeded by anything with a claim to being human. This threshold he set at 750 ml; and he noted that though the test was passed by Pithecanthropus, the australopithecines failed it. Piltdown, of course, passed too; but though its fraudulent nature had not yet been demonstrated, Piltdown had by this point become something of an embarrassment, and Keith was forced to shunt it into an evolutionary siding, as an "aberrant" form which had later become extinct. Interestingly, at about the same time in America the Harvard anthropologist E. A. Hooton

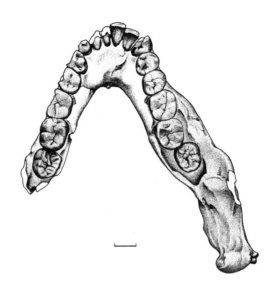

Slightly distorted lower jaw (TM Sts 53b) of Australopithecus africanus, *with complete but somewhat damaged dentition, from Sterkfontein Member 4, South Africa. Scale is 1 cm.*
DM.

was wrestling with the same problem and reached a more negative conclusion about the australopithecines because they "lacked the brain overgrowth that is specifically human and perhaps should be the ultimate criterion of a direct ancestral relationship to man of a Pliocene precursor. Because they lacked brains they remained apes, in spite of their humanoid teeth. Since the Australopithecinae died out in Africa, while the gorilla and chimpanzee survived, it would appear that a thorough-going ape is better than half a man". The logical flaw (or series of them) is glaring indeed; but perhaps it is redeemed by Hooton's poetical contribution to the argument over the australopithecines:

> Cried an angry she-ape from Transvaal
> Though old Doctor Broom had the gall
> To christen me Plesi-
> anthropus, it's easy
> To see I'm not human at all.

No energy need have been wasted on this empty argument if the various protagonists had taken a moment to consider how great a stumbling block the vagueness of the concept of "human" (and of "ape", come to that) is to the understanding of our evolutionary history. Until the idea of evolution arrived on the scene it was, of course, obvious what "human" and "ape" meant: living *Homo sapiens* with all its unique attributes on the one hand, and the chimpanzee and its like on the other. But with the idea of evolution came the notion of intermediates, that we are joined to the rest of the living world. And in this view the question with which Keith and others struggled seemed inevitably to arise: at what point did our precursors, less and less like ourselves as time recedes, become human? (We less often ponder the other side of the coin: what is apishness, and when did the ape precursors become apes?) But for better or for worse, it is as legitimate to use the adjective "human" in the inclusive sense of being related to us by descent as in the exclusive one of applying to creatures with the qualities that distinguish us from the rest of the living world. Clearly, these two senses of the word are in conflict: an early member of our own lineage need not have possessed any of the qualities of the mind which we see as unique to ourselves—and certainly could not have had all of them. Indeed, it need only have possessed one or two of the anatomical novelties that *Homo sapiens* has acquired over the long course of its evolutionary history. Most anthropologists today would lean toward the inclusive use of the term, to embrace the australopithecines as well as later fossil members of the human group: but it is important to remember that (depending on the characteristics that you regard as being typically "human") it might be difficult to view some of these members of the human family as "human" in a functional sense.

In any event, even as the outside world began to acclaim his discoveries, Broom was running into trouble back home in South Africa. Earlier in his career he had attracted the disapproval of his colleagues for supporting his paleontological research not simply by medical practice but by the sale of fossils to overseas institutions. Some rancor remained, and in the immediate postwar years this combined with concern for the ways in which Broom was recovering fossils from his cave sites. These localities were complex formations which had started life as subterranean cavities. They had filled up over long periods as debris of all kinds, including bones, fell in through shafts connecting with the surface. The stratigraphies within the cavities were clearly complicated, as were their subsequent histories of exposure at the surface due to erosion of the surrounding rocks; and in certain quarters it was felt that the use of dynamite to blast apart the hard cave breccia was not the method of fossil recovery most conducive to the precise stratigraphic control needed if the fossils were to be reliably dated by faunal association. This concern was fair enough—if a bit overpunctilious given that the consolidated sediments were rock-hard—and, indeed, some degree of uncertainty over dating still mars our knowledge of every South African australopithecine site, though this can hardly be regarded as Broom's fault.

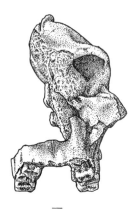
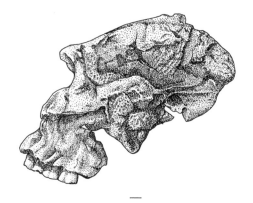

Front and side views of the cranium SK 48, from Swartkrans Member 1, South Africa, among the best-preserved remains of Paranthropus robustus *(or* P. crassidens*).*   DS.

But when the Historical Monuments Commission barred Broom from further work unless a "competent field geologist" participated, it was an egregious affront to one who had for years served as professor of geology and zoology at Victoria College in Stellenbosch. The outraged Broom redoubled his efforts in defiance of the commission, which ultimately relented. Within days, his efforts paid off with the recovery of a virtually complete, if edentulous, cranium at Sterkfontein, and by the end of 1947 Broom had amassed further specimens including a lower jaw with teeth and a partial skeleton with a more or less intact pelvis (which showed pretty conclusively that its possessor had been an upright biped) and much of the vertebral column. These finds were monographed in 1950, by which time little opposition remained to Broom's interpretation of them as "types of ancestral man just a little more primitive than Pithecanthropus."

With undiminished vigor Broom shifted his attention in 1948 to another site in the neighborhood of Sterkfontein, this one known as Swartkrans. This has ultimately turned out to be perhaps the most prolific site of its kind, and Broom and his assistant John Robinson had hardly started work there before they began to find early human fossils. The first of these to come to light was an adolescent mandible of a robustly built form reminiscent of the one from Kromdraai; in 1949 Broom named this specimen *Paranthropus crassidens*, in celebration of its large chewing teeth. Before his death in 1951 he had recovered several crania of this species in varying degrees of completeness. The 1949 season also yielded a couple of jaws representing a more lightly built hominid that Broom and Robinson dubbed

Telanthropus ("far man") *capensis*; they considered it to be "intermediate between one of the ape-men and true man," perhaps comparable to the Mauer individual from Germany. Robinson subsequently allocated these new fossils to *Homo erectus*, as Pithecanthropus and Sinanthropus had both become known by the early 1950s. By the time that Robinson concluded his investigation of the site in late 1952, the considerable hominid collection from Swartkrans also included postcranial remains, including some hand bones and part of a pelvis. A few crude quartz artifacts were recovered from dumps of material removed by earlier lime miners, but they attracted rather little attention until later work at the site by C. K. Brain revealed a substantial stone tool assemblage in association with hominid remains.

Dart meanwhile had shown little inclination to pursue further paleoan-thropological investigations—even though as early as 1925 fossils began to show up at other sites besides Taung. In that year a local schoolteacher, Wilfred Eitzman, had sent Dart some fossil bones from Makapansgat, a lime mine in the northern Transvaal. At the time Dart failed to follow up, but in 1945 a team of archaeologists began working at the nearby Cave of Hearths. Among those excavating there was James Kitching, later famous as a finder and describer of Karroo reptiles, and on his day off Kitching wandered over to the now-abandoned Makapansgat limeworks and began picking through the lime-miners' dumps of cave breccia fragments. Fossils soon showed up, and as a result Dart finally decided to launch a field program at Makapansgat. During 1946 and 1947 a team led by the Kitching brothers and Alun R. Hughes sorted laboriously through the breccia dumps. This work revealed another treasure trove of fossils, and thousands of mostly fragmentary bones were eventually recovered. Human fossils were not especially common among them*, but they did include a partial cranium, several jaws, and some postcranial bones. These appeared to be more lightly built than the robust remains from Kromdraai and Swartkrans, and Dart allocated them to the new species *Australopithecus prometheus*, named for the Greek hero who stole fire from the gods. He did this in the belief that the blackened condition of many of the bones from Makapansgat showed that these creatures had used fire, much as Davidson Black had surmised at Zhoukoudian. It later turned out, however, that the blackening was in fact due simply to the deposition of manganese on the bones.

More significantly, Dart's analysis of the bone fragments suggested to him that these early humans had wielded what he called the osteodonto-keratic (bone, tooth, and horn) culture. Dart based this conclusion on the

---

*Ron Clarke—of whom we'll hear more later—once figured out that perhaps as few as eight hominid individuals were represented among the huge numbers of fossil fragments recovered.

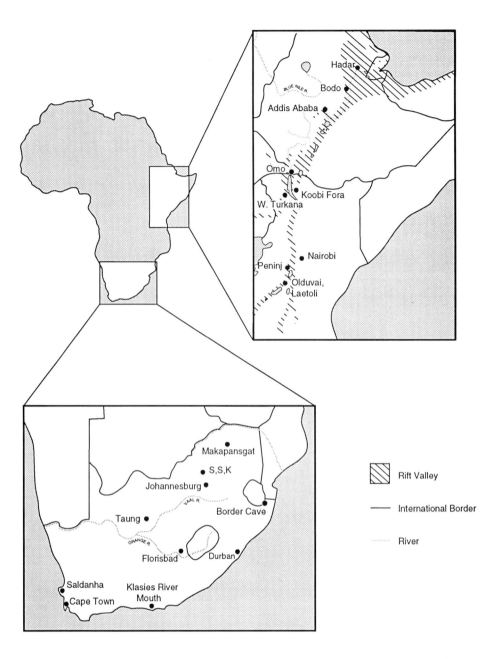

Map showing the locations of hominid-bearing sites in southern and eastern Africa, S,S,K indicates the location of Sterkfontein, Swartkrans, and Kromdraai.   DS.

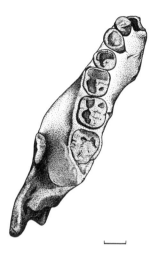

The MLD 40 half-mandible from Makapansgat Member 3, South Africa. Scale is 1 cm.

DM.

fragmentary condition of the mammal and other bones from Makapansgat and on the proportions of various body parts and animal species represented. He considered a number of possible explanations for the accumulation of bones within the cave, and he concluded that they could have found their way inside, and have been broken as they were, only as a result of the hunting and butchering activities of his *Australopithecus prometheus*. Many of the bone, tooth, and horn-core fragments, he believed, had been used as tools by these early humans, while the rest represented food remains. Further, Dart perceived similar patterns in the assemblages of animal bones recovered from Zhoukoudian and various European caves; and to his mind this strengthened both his specific theory about the Makapansgat bone accumulations and his more general conclusion that osteodontokeratic tool assemblages had been widely used by early hominids.

This was pioneering work, even though later studies showed it to be in error, and it enabled Dart to make various suggestions about australopithecine behavior. Some of these concerned the specifics of butchering techniques and dietary preference; other inferences were more wide-ranging, involving far broader aspects of intelligence, material culture and behavior. The frequency among the bones of shattered crania and hoofs, for example, showed to Dart the "focusing of australopithecine minds upon the sources of power," a human trait [that] can only be interpreted by the man-like comprehension of *A. prometheus* that the heads and feet of animals embodied the animals' strength; and that this strength could not only be extracted from the animal but be assimilated to their victors, and be turned back upon the animals themselves for their undoing. The intellectual ability and manual dexterity displayed by the performance of these feats formed the background of their promethean culture."

What's more, "to that osteodontokeratic psychological indoctrination humanity added stone and metal but has never been able to free itself phylogenetically or ontogenetically from bone, tooth and horn." Seeing evidence for cannibalism in the way his australopithecine fossils were fragmented, Dart went on in his later writings, much less measured than Weidenreich's on Zhoukoudian, to characterize these creatures as bludgeon-wielding "murderers and flesh hunters," whose violent proclivities led inevitably to the "blood-spattered, slaughter-gutted archives of human history." Stirring stuff—and it certainly caught the imagination of popularizers such as Robert Ardrey, who in *African Genesis* and other books painted a picture of the bloody birth of mankind as a willful predator unlike any other: a marauding "killer ape" whose vicious disposition has passed to us intact.

Dart's extravagant claims predictably aroused considerable criticism, and later studies have shown that the Makapansgat bone assemblages more plausibly result from the combined activities of water and porcupines than from the predatory behavior of australopithecines. Nonetheless, they did feed into a general tendency of the time to substitute a cultural concept of what constituted mankind for the anatomical one embodied in such notions as Arthur Keith's "cerebral Rubicon". Interestingly, anatomists were among the foremost proponents of the idea that evidence for "human" status was best sought in behavioral evidence, preferably stone tools. Weidenreich, for one, had as early as 1948 attacked Schepers's behavioral inferences from the brain casts of *Plesianthropus* with the declaration that "cultural objects are the only guide so far as spiritual life is concerned." Solly Zuckerman took a similar tack: what mattered was not so much what the brain looked like, but what it was used for. All the inferences in the world about behavior were for naught in the absence of evidence of toolmaking—and stone tools, the benchmark, were lacking at that point from the South African sites. Ironically, then, Dart's claim that these human precursors indulged in the characteristically human activity of toolmaking was coming under harsh criticism at the very moment when his insistence on the human, or at least prehuman, status of *Australopithecus* was becoming generally accepted.

Several thousand miles to the north, in East Africa, a shortage of stone tools was not Louis Leakey's problem. Born in Kenya to a missionary family, at the age of twelve Leakey was already collecting stone tools around his home near Nairobi. During his studies at Cambridge, he participated in a British Museum expedition to collect fossils in Tanganyika and thus added paleontological experience to his formal training in archaeology and anthropology. After graduating he led several archaeological expeditions to East Africa, gradually focusing his interest on Olduvai Gorge, a thirty-mile-long ravine cut into the Serengeti Plains in northern Tanzania. The story goes that this chasm, in places as much as 300 feet deep, was discovered in 1911 when a German entomologist named Kattwinkel almost fell into it as he chased butterflies across the Serengeti. Climbing down into the gorge, which cuts through many layers of sediments, Kattwinkel discovered large quantities of fossil bone lying around on the surface. The samples he took back to Germany caused a stir when it was found that they included parts of an early horse that had been extinct in Europe since the Pliocene, and a follow-up expedition was mounted in 1913 under the direction of Hans Reck, of the University of Berlin. Reck's team collected large numbers of fossils and drew up a preliminary geological map of the gorge, but the find that attracted most attention was that of a human skeleton.

Later sediments accumulate on top of earlier ones, so when a pile of sediments is exposed by erosion, as occurs in the walls of the Gorge, the lower layers represent an earlier time than the higher ones do. The skeleton was found fairly far down in the sequence of geological layers through which the Gorge was cut, and thus in sediments that were quite old, as the early Pleistocene mammal fossils found in the same layer seemed to testify. The anatomy of Reck's fossil human skeleton was modern, leading him to deduce that he had evidence for the very early existence of modern humans, much as Arthur Keith had lately concluded in the case of the Galley Hill skeleton. But, like Keith, Reck found himself assailed by colleagues who insisted that the skeleton was an intrusive burial into earlier deposits, and therefore testimony of no such thing. Clearly, further investigation was needed. But World War I intervened, the British took control of Tanganyika, and the plans of German anthropologists to follow up on this suggestive find were thwarted. It was not until Leakey took one of his expeditions to Olduvai in the early 1930s that research there resumed.

One of the odd things about Reck's collection from Olduvai was that while it contained the remains of a perfectly modern human, it was completely lacking in anything he could identify as a stone tool. Once Leakey arrived at the gorge it became apparent why. Stone tools were actually there in abundance, but they were not made of the flint to which European archaeologists were accustomed. Flint is a perfect material for stone toolmaking, because it fractures cleanly and predictably, producing sharp edges.

But at Olduvai, and in East Africa generally, flint is lacking. Although small quantities of the volcanic glass obsidian were available in places, early East Africans had generally to make do with coarser, more crystalline materials. Such rocks—basalt, quartzite, and so forth—produce cruder, less fine-looking tools than flint, and the Germans had simply failed to recognize various unimpressive lumps of rock scattered around Olduvai for the tools that they were. With his eye locally trained, Leakey had no such problem. Arriving with Reck in 1931, he almost at once began to find tools. And just as Leakey had no trouble in persuading Reck that what he had found were indeed crude stone tools, once on the spot Reck seems to have had equally little difficulty in persuading the initially skeptical Leakey of the ancient-ness of his 1913 skeleton. Indeed, the presence of tools seems to have been the critical factor that changed his mind, and Leakey went on to document the occurrence of stone artifacts in all Reck's four major stratigraphic units of the gorge, from Bed I at the bottom to Bed IV near the top. These, Leakey believed, showed a progressive evolution from crude "pebble tools" at the bottom (belonging to a culture he later termed the "Oldowan") to handaxes at the top. Given the differences in raw material, these latter were remark-ably similar to the handaxes of the Acheulean tradition Boucher de Perthes had recognized in France.

Later in the same year Leakey visited a couple of adjacent sites in western Kenya called Kanam and Kanjera. At the latter he found some unimpressive scraps of modern-looking braincase, and at the former the front part of a rather battered lower jaw which he also assigned to *Homo sapiens* and which he claimed was older than the Olduvai skeleton. Back in England Leakey found his audience remarkably receptive to this ambitious though hardly well-substantiated contention, but it was not long before a distinguished geologist once more disputed the claim of great antiquity for the Olduvai skeleton. Leakey took this critic, Percy Boswell, with him on his next East African trip, in 1935, and to his great embarrassment found himself unable to substantiate his case. Nonetheless, he continued work at Olduvai, iden-tifying a large number of artifact-yielding sites, making geological obser-vations, and collecting large numbers of vertebrate fossils. At the end of the season, acting on a tip from a local Maasai, he visited some exposures nearby (relatively speaking; the trip from Olduvai took Leakey three days, though it's now about ninety minutes) at a place called Laetoli, and he made a small collection of fossils there. These were believed at the time to be of about the same age as Beds I and II at Olduvai, and they included a hominid canine tooth which remained unrecognized as such until 1981. The area was revisited in 1939 by the German explorer Ludwig Kohl-Larsen, who made a large collection of fossils that were unfortunately derived in-discriminately from deposits of different ages. Among these fossils were a

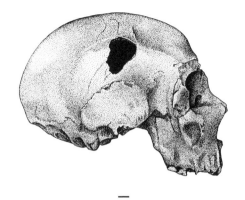
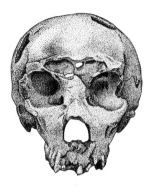

—   —

Side and front views of the Neanderthal cranium from Saccopastore, Italy. Scales are 1 cm.   DM.

small piece of hominid upper jaw, containing two teeth and an isolated molar, that later served the paleontologist Hans Weinert as the basis for a new species, *Meganthropus africanus*. *Meganthropus* was a genus that had been set up by von Koenigswald to contain a massive fragment of lower jaw that had been found in the Sangiran deposits of Java. Laetoli was to figure importantly later on in the story of human evolution, but Leakey did not visit the area again until 1959, when remarkable news was about to emerge from Olduvai.

But back in 1935 the glory days of Olduvai were still a quarter-century in the future. The reputations of whiz-kids are notoriously fragile, and Leakey's was severely dented by the debacle of the Olduvai skeleton, which was indeed a later burial into earlier deposits (to be precise, into the top of Bed II, after the overlying Beds III and IV had been eroded away). This, combined with an acrimonious divorce in 1936, placed an academic appointment in England beyond his reach, and from then on Leakey based himself in Kenya, working across the border at Olduvai as time and finances allowed.

Elsewhere in the world, meanwhile, other finds were sporadically coming to light. In Europe, for instance, the two decades leading up to World War II saw the recovery of a number of new Neanderthal-like fossils. In 1926 the British archaeologist Dorothy Garrod recovered a juvenile Neanderthal skull from the cave of Devil's Tower in Gibraltar, and three years later a gravel quarry at Saccopastore, near Rome, yielded a quite lightly built Neanderthal skull that seemed to date from the last interglacial. It was accompanied by a variant of the Mousterian industry similar to that found at the

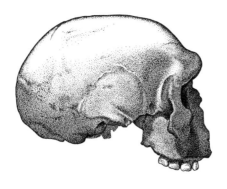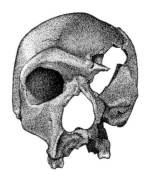

—   —

Side and front views of the distorted cranium from Steinheim, Germany. Scales are 1 cm.

DM

somewhat later Italian site of Monte Circeo, where in 1939 a more typical Neanderthal skull was found in the center of a stone circle. Further afield, at about the same time, a partial skeleton of a nine-year-old Neanderthal boy was found at Teshik-Tash in Uzbekistan, apparently buried within a circle of goat skulls. Dating of this specimen is still hazy, but it remains important in showing how far to the east humans of typical Neanderthal morphology spread during the later part of the Pleistocene.

Less easy to interpret were the Swanscombe (England) and Steinheim (Germany) crania discovered in the mid-1930s. The latter was found in river gravels dated faunally to the second interglacial period (Mindel-Riss), hence earlier than most if not all Neanderthal sites (though the original estimates placed it later). The specimen itself consisted of a reasonably complete if somewhat distorted cranium which was said to combine Neanderthal with more modern traits, despite a low estimated brain volume, at about 1,100 ml. Neanderthal brain volumes, by contrast, were coming in regularly at about the modern average, or possibly even a bit more. Arthur Keith suggested that the affinities of the Steinheim skull lay with the Ehringsdorf specimen described by Weidenreich, which not all were happy to accept as a Neanderthal. The Swanscombe fragments seemed to be about the same age, and initially consisted of the occipital bone from the back of the skull and the associated parietal bone forming the left side; the right parietal was found twenty years later! Most judged at the time that the part of the skull preserved was remarkably modern-looking, and the estimated brain capacity, at 1,325 ml, was well within the modern range. Keith declared this specimen to be a precursor of Piltdown, while others discovered more Neanderthal-like characteristics in the back of the skull.

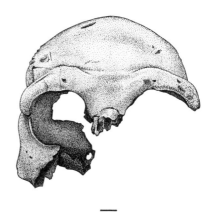

Side and front views of the facial fragment from Zuttiyeh, Israel ("Galilee Man"). Scales are 1 cm. DM.

Even more interesting and suggestive discoveries were being made in the Levant at about the same time. In 1925 a frontal bone with marked brow ridges but a relatively high forehead was exhumed by the English amateur Turville Petre, along with a couple of other fragments of what may well have been a complete skull, from the cave of Zuttiyeh in Palestine (now Israel). It was found in association with a "Levalloiso-Mousterian" stone tool industry very similar to that of the European Neanderthals. As it happened, flint tools of this kind were also being picked up by visitors to the cave of Qafzeh, only a few miles away outside Nazareth. In the early 1930s these piqued the interest of Réné Neuville, the French consul in Jerusalem, who began excavations in 1933. He rapidly uncovered an archaeological sequence that spanned the Middle and Upper Paleolithic. In the Middle Paleolithic (Mousterian) strata Neuville and his colleague Moshe Stekelis found fragments of five human skeletons, all of them anatomically modern. These were provisionally dated to the early part of the last glacial, but during Neuville's lifetime they remained undescribed, failing to attract the attention paid to those from the Mount Carmel sites (see below). Indeed, it is only very recently, with advances in dating techniques, that Qafzeh's full importance has been realized.

Much more influential at the time were excavations under way in limestone caves at the western foot of Palestine's Mount Carmel, in sight of the Mediterranean. Between 1929 and 1934 a team under the direction of Dorothy Garrod excavated the cave of Mugharet et-Tabūn, finding an almost complete female human skeleton and a more robust lower jaw interpreted

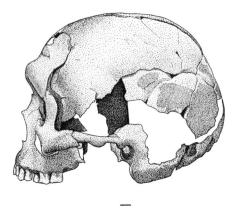
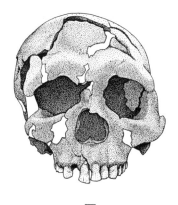

Side and front views of cranium 6 from Jebel Qafzeh, Israel. Scales are 1 cm.   DM.

as male. In the same layer were found numerous mammal bones, which appeared to date the humans to some time in the last interglacial period. While the female skull was more lightly built than those of most western European Neanderthals, and was also more rounded in the occipital (back) region, it otherwise possessed distinctly Neanderthal features. Once more, the associated lithic culture was Levalloiso-Mousterian. Nearby, in 1932 the same team excavated a virtual cemetery in the rock shelter of Mugharet es-Skhūl, where remains of at least ten individuals of quite modern morphol-

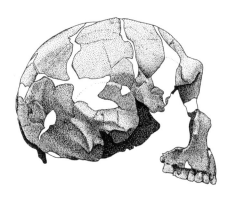
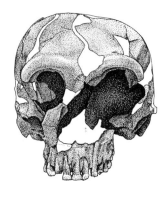

Side and front views of the cranium C1 from Tabūn, Israel. Scales are 1 cm.   DM.

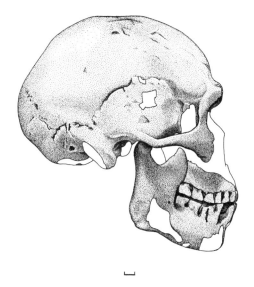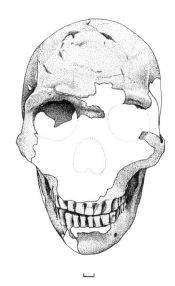

Side and front views of the most complete human skull from Skhūl, Israel (Skhūl V). Scales are 1 cm.

DM.

ogy were exhumed. The individuals of this population are unusual principally in having distinct if small brow ridges, but this has not prevented most recent authors from regarding them as essentially modern. Again, the stone tools from the site were Levalloiso-Mousterian, and despite certain faunal differences it was believed that the Tabūn and Skhūl populations were more or less contemporaneous, dating to the last interglacial.

The fossils from the two Mount Carmel sites were monographed together in 1939 by Theodore McCown and Arthur Keith, who described them as representing a single variable population. As good morphologists they clearly had difficulty in arriving at this conclusion, especially since the only anatomical basis they could find for it lay in similarities of the dentition. McCown and Keith felt obliged to group these fossils principally because they shared similar lithic industries, and because the two sites were only a stone's throw from each other and, they believed, of about the same age. This made it necessary to explain why a population containing humans as different as those from Tabūn and Skhūl should be so extraordinarily variable. Were these people "in the throes of an evolutionary transition and therefore unstable and plastic in their genetic constitution"? Or was "the variability due to hybridity, a mingling of two diverse peoples or races"? McCown and Keith preferred the former explanation, though others have leaned toward the latter. In any event, by evoking these two possibilities McCown and Keith set the pattern of discussion of the Mount Carmel people for years to come.

In comparing the Mount Carmel assemblage with others McCown and Keith clearly faced a difficult situation, lacking as they did a homogeneous assemblage. They saw a spectrum of late Pleistocene fossil humans with one extreme represented by the European Neanderthals and the other by the Cro-Magnons. In the middle lay Mount Carmel, with the Tabūn types closer to the former, those from Skhūl to the latter. Between the Tabūn variety and the "classic" Neanderthals of western Europe lay the Krapina people, representing a more lightly built intermediate variant, while the Skhūl and Cro-Magnon forms were considered very close indeed to each other, both showing specifically European characteristics. But instead of concluding from this perceived continuity of form that Neanderthals had simply evolved into modern Europeans via a Mount Carmel-like stage, McCown and Keith proposed a more complex scenario. Noting that Neanderthals were ubiquitous in western Europe during the "middle Pleistocene," but that this type became progressively modified from west to east until in Palestine there appeared a type transitional toward the modern, they extrapolated this trend to suggest that the ancestors of the modern European population had arisen in western Asia somewhere to the east of Palestine. The Mount Carmel people themselves were therefore "not the actual ancestors of the Cromagnons but Neanderthaloid collaterals or cousins of that type." It took me several readings of McCown and Keith's breathtakingly opaque presentation to grasp this reasoning. And this lack of crystal clarity provides, I think, a good example of what happens when you allow factors other than morphology to hold sway when making decisions as to evolutionary relationship. It also, it must be admitted, reflects the fact that McCown and Keith didn't exactly see eye to eye. Weidenreich, though, was delighted by McCown and Keith's declaration that "Eastern Asia [was] the cradle of the proto-Mongols," and as a reciprocal courtesy added Palestine to his list of centers of the evolution of modern humans, the Levant becoming the source of modern Europeans.

7

The Synthesis

Although you don't have to wander far if at all beyond the confines of the order Primates to demonstrate virtually the full spectrum of evolutionary phenomena, it remains true that anthropology has contributed rather little to the general development of evolutionary and systematic theory. Perhaps it's not entirely surprising, then, that it took a while for paleoanthropology to feel the effects of a profound transformation of evolutionary thought that took place during the 1930s and 1940s. During these years Broom could on the one hand agree with Alfred Russel Wallace that "the large brain could not have arisen by Natural Selection" and von Koenigswald could on the other insist that size increase was inevitable, while neither claim sounded particularly odd to anthropological ears. Yet well before these sages wrote the foundations had already been laid of what was rapidly (except, perhaps, in France) to become an evolutionary orthodoxy that has lasted up to our own day.

By the 1920s and early 1930s the fragmentation of evolutionary biology to which I've already referred had become downright chaotic. Almost everyone had his own theory of evolution, and few of these were darwinian in the sense that natural selection was accepted as the prime force behind evolutionary change. Antiselectionism was, indeed, the order of the day. Many biologists, including anthropologists such as Weidenreich, were still infected to one extent or another by the concept of orthogenesis, the idea that evolution was directed by an inner drive toward a fixed goal. A variant of this idea held that evolutionary change was the expression of some kind of innate potential that lay within each lineage of organisms. Saltationism was still rife, whereby new kinds of organisms were seen as arising abruptly by some mechanism of discontinuous variation. "Mutations" were often viewed not as the small genic alterations to which we refer by this term today, but as quantum changes giving rise to new species; it was, indeed, in this sense that Hugo de Vries had coined the term back at the beginning of the century. And where mutations were spoken of in the modern sense, it was often concluded that "mutation pressure" was the driving force of evolution. Various forms of Lamarckian thought were also current, among them the ideas that environmental changes somehow induced changes in the organism, or that use or disuse of organs during the life of an individual controlled how those organs would appear in its offspring. Even some avowed darwinians incorporated elements of such thinking in their evolutionary theories, combining natural selection with ideas of "soft"—broadly, Lamarckian—inheritance. Weissmann's demonstration that the germ line (genes) and soma (body) were separate things had yet to become accepted throughout biology, although decades before it had instantly been absorbed by Wallace and several others to produce a "neo-darwinian" school of thought.

Moreover, as the ornithologist Ernst Mayr has chronicled in several fascinating historical accounts, the first third of the twentieth century witnessed divergence to the point of alienation between the varied disciplines which we think of today as parallel branches of evolutionary biology. The geneticists pursued their genes, often using fruit flies as their experimental models; the zoologists sought the nature of species and the mechanisms of their origin; and the paleontologists applied themselves to the description and classification of the species of which the rapidly growing fossil record was composed. Between these various specialties, then, there rapidly developed some pretty massive barriers of mutual incomprehension. In 1944 the paleontologist George Gaylord Simpson, Mayr's colleague at the American Museum of Natural History, summed up this situation in a famous passage:

> Not long ago, paleontologists felt that a geneticist was a person who shut himself in a room, pulled down the shades, watched small flies disporting themselves in milk bottles, and thought that he was studying nature. A pursuit so removed from the realities of life, they said, had no significance for the true biologist. On the other hand, the geneticists said that paleontology had no further contributions to make to biology, that its only point had been the completed demonstration of the truth of evolution, and that it was a subject too purely descriptive to merit the name "science".

Perhaps such attitudes resulted almost inevitably from what seemed to be rather fundamental differences in preoccupation. Apart from a rearguard action against the waning advocates of soft and "blending" inheritance, the stuff of the genetics of the time was the study of how the frequencies of different alleles within populations might change given mutation both at the level of the gene and at that of the chromosome, where it was coming to light that mix-ups involving whole strings of genes could also occur. And there was no evident relationship between gene frequency (or "microevolutionary") change and the "macroevolutionary" phenomena that were of primary interest to naturalists: the origin of species, the existence of higher taxa*, evolutionary diversity, and so forth. Mayr explained the problem not long ago by suggesting that in fact all these elements actually belong to a single hierarchy that runs from molecules and genes, through individuals, populations and species, to higher taxa and associated macroevolutionary phenomena. The difficulties arose at least in part because the geneticists were looking at the lower levels of the hierarchy while the naturalists were interested in the higher ones. But in part only: the rest of the confusion lay in such things as the tenacity of saltationism and the refusal of soft inheritance to die, and these were factors that also did much to ensure that even within each major field of study discord was the order of the day.

Hardly fertile ground, on the face of it, for agreement within disciplines, let alone among them; yet by the mid-1940s a "synthesis" had been achieved in evolutionary thought to which almost everyone subscribed. It is still a little unclear to me exactly how this synthesis came about, or at least how it so rapidly gained acceptance; but it involved the integration of darwinian ideas of natural selection with changing frequencies of genes in populations, and perhaps the neatest early demonstration of how this might be done was the idea of the "adaptive landscape" first put forward by the American geneticist Sewall Wright in 1932. As we've seen, the hereditary material—the genes—is strung out along the chromosomes. Each gene occupies a particular spot, or "locus," on its chromosome, at which it may be represented by one of several different alleles, or alternative forms. With many thousands of loci, each with several alleles, the "gene pool" of any population contains vast numbers of different combinations of alleles, or "genotypes." Some of these, surmised Wright, had to be more favorable than others, producing "fitter" individuals better able to survive and reproduce in any given environment. He drew the equivalent of a topographical map in which the contour lines connecting areas of equal elevation were replaced by lines delineating regions of fitness. On the hilltops were clustered the fitter genotypes, while the less fit genotypes occupied the valleys. On this analogy, the major problem for each species was to maximize the

---

*A *taxon* (plural: *taxa*) is a named unit at any level of the hierarchy of classification (species, genus, family, order, and so forth). *Higher taxa* are those above the level of the species.

number of individuals occupying the hilltops and to populate the valleys as little as possible. This was a powerful analogy, and essentially a darwinian one, even though Wright recognized that chance factors ("genetic drift") might also affect the survival of new alleles or gene combinations. The relative fitness of individuals, central to Darwin's ideas, virtually implied natural selection.

The adaptive landscape metaphor was rapidly picked up on and elaborated by many different people, often way beyond Wright's original intentions. The peaks, particularly, came to represent many different things. But however it is interpreted this analogy clearly marries selection and gene frequencies, and it is this combination which set the scene for developments to come. Many different scientists contributed to the emerging "synthesis" in evolutionary thought, but the major coherent statements of its principles appeared in three books. The first of these, the geneticist Theodosius Dobzhansky's *Genetics and the Origin of Species*, appeared in 1937; Mayr followed up in 1942 from the systematist's viewpoint with *Systematics and the Origin of Species*; and Simpson brought paleontology into the fold in 1944 with *Tempo and Mode in Evolution*. Each of these works obviously had its own particular focus, but each accepted a few simple basic principles. First, evolution was a gradual, long-term process, essentially consisting of the accumulation within lineages of small genetic mutations and recombinations. Over enough time, the accumulation of minor changes would result in large effects. Second, this generation-to-generation change was controlled by natural selection, environmental factors promoting adaptation within the lineage via the differential reproductive success or failure of different variants. As environments changed, populations would change to keep in step and maintain or improve their adaptedness. Third, this same process of the gradual accretion of genetic (hence physical) change could be extrapolated to explain higher-level phenomena, such as the origin of new species and of biotic diversity.

It was agreement on this last point that was crucial in bringing together those who studied genes and those whose central interest was organisms. For while the gradual accretion of genetic change implied a basic continuity, it was nonetheless evident to anyone concerned with the world of living organisms that it was marked by discontinuities: each species in the marvelously varied living biota was an isolated genetic package, distinct from all others. If the basic mechanism of evolution was simply one of small changes summing up over time to produce major effects, where did these discontinuities come from? The adaptive landscape had the answer. Its peaks were favorable ecological niches, to which their occupying populations were adapted. In contrast, the valleys between were hostile areas and no individual of a species could afford to slide too far down the slope toward the valley floor. But the map didn't stay constant: as Simpson put it, it was "more like a choppy sea than a static landscape." Natural selection had to operate incessantly to keep each population nicely balanced on the peak that was shifting beneath it, and from time to time a peak itself would divide, giving rise to two peaks, which then moved apart. With natural selection working in different directions on each new peak, in time the occupying populations, once one, would have diverged enough to be separate species. Presto! Speciation—the production of new species—was reduced to another aspect of adaptation, itself founded on the slow accumulation of tiny genetic changes. And repetition of this simple process over long periods of time would ultimately give rise to new genera, families, orders, and on up. Among mammals, at least, this meant that speciation was unlikely to occur without some external barrier (a seaway, a desert, a mountain range) dividing a widespread population into two. But once that had occurred, the result was almost a foregone conclusion.

This was, of course, great for the geneticists, and particularly for the mathematical modelers among them. For it meant that the key to evolutionary change was entirely held by the guardians of the nascent and model-oriented science of population genetics. It was also fine for the zoologists who, while having to concede that species were not discrete packages in time, were still able to view them as discrete in space, which was what counted in their day-to-day business. But it was awkward for the paleontologists. It is tough to be a scientist without a basic conceptual unit of study, and the synthesis had robbed paleontologists of this fundamental necessity. For it is certainly not provided by the individual fossil (unless that fossil happens to be the only one of its kind that you have). The first thing you need to know about any fossil, before you can fit it into any larger picture, is, "what species does it belong to?" And under the synthesis, species inexorably evolve themselves into other species, by constant tiny incremental changes. Over time, which is the unique property of the fossil record, species lose their identity: it is impossible to say where any one starts or finishes. Lineages, ancestor-descendant sequences of populations in the fossil record, may undergo enormous quantities of change from their beginnings to their ends. And any member of that sequence *must* have belonged to a particular species. But lineages under this viewpoint can be divided up only in an arbitrary manner; and beyond depriving paleontology of a theoretically rigorous structure, on a practical level this creates the potential for unresolvable disagreement. It is for this latter reason that post-synthesis times witnessed the remarkable spectacle of paleontologists congratulating themselves on the deficiencies of their data base; the famous gaps in the fossil record, it was said, provided handy places at which to break up lineages into segments that could conveniently be labeled with species names in the same way as is done with living forms.

Small wonder, then, that rethinking of the claims of the synthesis to be a comprehensive explanation of the evolutionary process was ultimately to come from students of the fossil record. But there is no question that the architects of the synthesis had produced a magisterial achievement that swept away a huge panoply of evolutionary mythology. Gone were ideas like orthogenesis and saltationism, replaced definitively by Darwin's original concept of evolution as an opportunistic, non-goal-oriented process. In no field was this development more salutary than in the study of human evolution, where the temptation to produce orthogenetic or finalistic interpretations had been particularly strong, and where reluctance to abandon them was commensurately marked. And the triumvirate of the synthesis showed no reluctance to share the benefits of their insight with their paleoanthropological colleagues.

The synthesis had been born of the emergence of what Mayr has called "population thinking." Many early geneticists thought of species in terms of sets of particular intrinsic qualities, while systematists—those concerned with the description and analysis of the diversity of life—had traditionally been prone to regard species as types. It was recognized that species consisted of large numbers of individuals, but each individual was thought to conform more or less to a basic archetype. Population thinking, in contrast, involved the realization that species consisted of clusters of unique individuals and populations, and that there was no ideal "type" against which any individual could be measured. This opened the door to a new view in which local populations played a critical role in the production of new species and new adaptations. Most species are "polytypic," composed of several such populations, each one an aggregate of unique individuals and with its own geographical range. Each, too, differs slightly on average from the next, so that distinctive populations might not represent the separate species that a typologist would recognize. The critical test is not external form, but reproductive continuity. The synthesis recognized such geographically varying local races, or subspecies, as the engines of evolutionary change.

As early as 1944 Dobzhansky applied this kind of thinking to the human fossil record and concluded that "the differences between Peking and Java men are easily within the magnitude range of the differences between the living human races." Further, rejecting the conclusion of McCown and Keith, he found that the variability among the Mount Carmel fossil humans must have been due to hybridization between Neanderthals and moderns, two subspecies that had arisen elsewhere but which, belonging to the same species, had interbred on encountering one another. Ignoring a certain amount of circularity in his reasoning, Dobzhansky declared that the Mount Carmel people showed "how rash are the assertions of some writers that in fossil forms evidence on presence or absence of reproductive isolation can never be obtained!" From this he proceeded to the conclusion that, "in Hominidae, a morphological gap as great as that between the Neanderthal and the modern may occur between races rather than species," although it remained for him an open question whether Peking-Java and the Neanderthals were distinct species or only racially separate.

Dobzhansky then discussed what he considered the two leading current models of human evolution. On the one hand there was the "classic" view, which "produced a tree with many branches . . . [of which] the known fossils represent . . . only rarely the main phylogenetic trunk"; on the other there was Weidenreich's "parallel development of races," whereby several distinct racial lineages separately passed through the same general series of evolutionary stages to achieve modern form. Finding (unsurprisingly, given the amount of intraspecific variation he was prepared to accept) that "as far as [is] known no more than one hominid species existed at any one time level", Dobzhansky envisioned a highly complex set of local developments and hybridizations among human populations throughout the Pleistocene. But he concluded that the differences between the two hypotheses before him were hardly significant, given that pretty much the entire set of human evolutionary developments since Java Man had taken place within the confines of a single polytypic species. And although he was undoubtedly right to attack the typological approach to classifying fossils in the human record, in deliberately brushing complex events beneath the single species rug Dobzhansky was fostering what may be the most destructive canard in the entire long catalog of paleoanthropological misconceptions.

Not that his fellow synthecist Ernst Mayr had any objections. At an influential conference held at Long Island's Cold Spring Harbor in 1950 (at which Simpson still felt compelled to reiterate the arguments against such anachronistic beliefs as orthogenesis and the inevitability of size increase over time) Mayr tried to bring the naming and classification of fossil humans into what he saw as equivalence with those employed in other areas of zoology. Blown up to human size the six hundred species of *Drosophila* fruit flies would, Mayr claimed, look very much more different from each other than do humans and gorillas, or even monkeys. The fact that many distinctive populations formerly classified as separate species were now recognized as mere racial variants meant that we had to adjust our categories all the way along the line. While the chimpanzee and gorilla were clearly good species, Mayr said, they equally clearly did not deserve to be classified in separate genera. Similarly, there was no justification for separating the apes into a family (Pongidae) distinct from that of humans (Hominidae); and as for the australopithecines, well, they all belonged to the same genus. And this genus might, indeed, even be *Homo* since upright posture (which freed the hands, stimulating development of the brain) had been achieved. In the end, Mayr plumped for placing all known fossil humans in the genus *Homo*, within which he recognized the three species *H. transvaalensis*, *H. erectus*, and *H. sapiens* (including Neanderthals).

Like Dobzhansky, Mayr could find no evidence for the existence of more than one human species at the same point in time, despite accepting the about-to-be-exposed Piltdown remains (which Dobzhansky had dismissed as a fortuitous association of human and ape fossils) as a genuine combination. He explained loftily that the difficulties Weidenreich and others had experienced in reaching the same conclusion were simply due to their inability to comprehend that "one type does not change into another type evenly and harmoniously, but that some features run way ahead of the others." This rash assertion did nothing to diminish the influence of Mayr's analysis, and his paper ushered in among paleoanthropologists, if haltingly at first, an era of "lumping" (minimizing the number of taxa recognized), which was sharply at variance with the older "splitting" tradition derived from typology.

During the 1950s new technologies were filtering into paleoanthropology as well as new ideas on how evolution might be expected to occur. One of these was fluorine analysis, a technique first floated in the mid-nineteenth century but only seriously applied a hundred years later by Kenneth Oakley at the British Museum (Natural History). Fossils take up fluorine from the surrounding deposits at a steady rate. Thus, in theory, if you know how much fluorine there is in a fossil, you will also know how long it has been buried. In practice it turns out that fluorine content is affected by too many variables to provide a reliable general dating system. However, fossils that have been buried in the same place for the same length of time should indeed contain the same concentration of fluorine. And this obviously provides a means for determining whether two specimens found in the same stratum are contemporaneous. It was the application of fluorine analysis to the Piltdown specimens that precipitated the exposure of the fraud: the fossils of extinct mammals that had accompanied *Eoanthropus* contained high levels of fluorine, up to about 3 percent, while specimens attributed to the human form averaged an insignificant 0.02 percent. But perhaps as important, fluorine analysis and similar methods proved once and for all that Galley Hill and other putatively ancient humans of modern anatomy represented intrusive burials into the deposits from which they had been recovered.

Thus, at midcentury, a clearing-out process was in progress that swept away not only a large accumulated burden of discredited evolutionary theories but a whole suite of pseudofossils whose presumed great age had made nonsense of rational efforts to understand the human fossil record. One of the first paleoanthropologists to capitalize on these new developments was F. Clark Howell, then of the University of Chicago, who reanalyzed the Neanderthals in 1951. With such awkward forms as Piltdown and Galley Hill eliminated from consideration, he was able to envisage a single lineage in Europe that led from Mauer through Swanscombe to Steinheim and thence to an "early Neanderthal" assemblage in which such forms as Ehringsdorf and Saccopastore belonged. These forms seemed to date from the last (Riss-Würm) interglacial and to have shorter, higher, generally more lightly built skulls—hence more like modern humans—than the later "classic" Neanderthals (Neanderthal, La Chapelle-aux-Saints, La Ferrassie, Monte Circeo, et al.) of the last (Würm) glacial. Howell also saw a geographical trend, the early Neanderthals from sites in western Europe tending more toward the "classic" condition than those from further east (Krapina, Mount Carmel). From these distributions in time and space flowed the idea that early Neanderthals in eastern Europe and the Levant had given rise, via the Mount Carmel assemblage, to modern humans; toward the west, in contrast, slightly different early Neanderthals had given rise to the "classic" forms. Howell rejected the possibility of hybridization between classic Neanderthals and moderns because the evidence pointed so clearly to abrupt replacement in western Europe of the former by the latter.

This scenario made excellent sense in terms both of the dictates of the synthesis and the increasingly detailed record of Pleistocene glaciation that geologists were compiling in Europe and western Asia. During the mild climate of the Riss-Würm interglacial an early Neanderthal population was widespread across this region, with members of its western portion foreshadowing more strongly than those to the east the physical peculiarities of the "classic" form. With cooling of the climate as the Würm glaciation began, the western population became effectively cut off from that to the east as the Scandinavian, Alpine, and Pyrenean ice sheets expanded and converged, and conditions deteriorated in the corridors between them. Subjected to a harsher climate than their relatives to the south and east, and consequently to a severe regime of natural selection that was perhaps aided by genetic drift in small populations, the more or less isolated Neanderthals of western Europe acquired their exaggerated "classic" specializations.

In 1952 Howell looked more closely at the faunas accompanying the various classic Neanderthal finds and concluded that to the extent to which it was possible to judge, all were approximately contemporaneous, dating from the first part of the Würm glacial. Only with the beginning of an "interstadial"—a period of warmer climate within the glacial—did archaeological and later fossil evidence begin to appear of the occupation of western Europe by modern humans. By this time the Neanderthals had vanished, perhaps with the disappearance of the environment to which they had so closely adapted, or maybe because they were "extinguished" by their modern cousins as they arrived from an eastern place of origin. In some way I find hard to define exactly, but which probably has much to do with their infusion by the spirit of the synthesis and the discounting of Piltdown et al., these two papers by Howell seem to me to have inaugurated what one might call the "modern" era of paleoanthropological studies.

Another measure of modernity was lacking, however. As this century passed its midpoint, there were still no practical ways of determining the absolute ages—ages in years—of fossils. Dating was relative: this geological layer underlies this other one, therefore it is older; these extinct animals are found in older layers, while these others are found in younger ones; that site is older than another because it contains an older fauna. And although the labors of geologists and paleontologists over the decades had produced remarkably detailed local and worldwide relative chronologies, there was still no way to calibrate them in terms of the passage of years. A sort of calibration had been tried in terms of sedimentation rates—the rates at which sedimentary rocks are laid down—which is why many early evolutionary trees incorporated the depths of sediment characteristic of various periods much as we use time scales today; but it took the arrival of methods of chronometric dating (in years) to show just how approximate and unreliable such extrapolations were.

The first such method of dating to be introduced was the radiocarbon (carbon-14) technique, invented by Willard F. Libby in 1950. This, like most later methods, depends on the phenomenon of radioactivity, which is why such dating is often known as "radiometric". Many naturally occurring atoms (the radioactive ones) possess unstable nuclei that spontaneously "decay" to stable states of lower energy. When a radioactive "parent" atom decays, it changes to another type of atom known as the "daughter" product. The rate of decay is characteristic of the particular kind of atom involved, and is effectively independent of external conditions. In theory the time that it will take for all parent atoms in a system to decay is infinite, so rates of radioactive decay are expressed in terms of a "half-life", or the time it takes for half the atoms in a system to decay. The radiocarbon technique depends on the decay of carbon-14, an unstable form of carbon that is incoporated as a small but constant percentage of the carbon that all living things contain. When an organism dies its carbon-14 ceases to be renewed and begins to decline by decay as a proportion of total carbon present. Measuring the amount of radiocarbon relative to other carbon contained in an organic sample thus provides a way of finding out how much time has elapsed since the death of the organism concerned. At 5,730 years the half-life of radiocarbon is relatively short, so that when a sample is older than about 40,000–50,000 years (40–50 kyr) it contains an unmeasurably small amount of radiocarbon. This obviously places an effective limit on the usefulness of the method, although during the early days hopes were high that isotope enrichment techniques would extend its range by a few tens of thousands of years, and several very ancient dates, generally viewed with suspicion, were published. Another limitation, at least until recently, has been the size of the sample needed for effective measurement: besides considerable technical difficulties in using most fossil bone, destruction of a large chunk of it was necessary to get a date. Obviously, people didn't want to lose their precious human fossils in the process of finding out how old they were, so radiocarbon dates were, and generally continue to be, obtained on organic materials associated in the same deposits as the human fossils. Charcoal is a favorite.

Given the 40 kyr maximum of radiocarbon dating, this method was clearly going to be restricted to the latest part of the Pleistocene. Early results thus didn't shock paleoanthropologists, who were already quite confident that they were dealing in this period only with tens of thousands of years. But the new precision that radiocarbon permitted was exquisite, and archaeologists, particularly, were entranced. One of the earliest of these to capitalize on this new tool was Harvard's Hallam Movius. In 1954 he obtained a date of around 24 kyr for Gravettian layers at the Abri Pataud, a rockshelter in Les Eyzies which he was then excavating; before long he was able to propose a rough time scale for the latter part of the last glacial in western Europe. Mousterian artifact assemblages, more or less exclusively associated in that part of the world with Neanderthal physical types, started well outside the range of the radiocarbon technique, and persisted up to about 32 kyr. What is now known as the Châtelperronian industry, variously interpreted as the work of late Neanderthals or early moderns, began somewhat earlier than this and ran to about 30 kyr. The Aurignacian, the first undisputed Upper Paleolithic industry, began at around 32 kyr, while the end of the Pleistocene, which coincided broadly with the disappearance of the high cultures of the Upper Paleolithic, could be placed at around 10 kyr. Parenthetically, one might note that there is a growing tendency nowadays to conclude that in conventional geological terms the Pleistocene has not actually ended, and that we are now in fact simply in the most recent stage of the late Pleistocene.

In any event, it was clear early on that nonmodern human fossils were

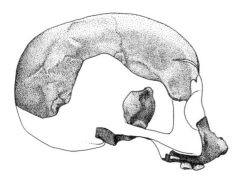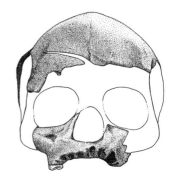

Side and front views of the early modern human skull from Niah Cave, Sarawak. Scales are 1 cm. DM.

likely to be found only towards the outer limit of the radiocarbon technique; and additionally, because sites yielding human fossils are much rarer than sites with artifacts, dates associated with actual human fossils were generally slow to come in—although a calibrated archaeological record went a long way towards providing the needed time scale for the latest phase of human evolution. Perhaps the first dates for fossil humans were obtained in 1951 by Junius Bird of the American Museum of Natural History for specimens from South America; but these were post-Pleistocene, if only just, and dates directly associated with nonmodern humans began to accumulate significantly only during the 1960s. Perhaps the most intriguing radiocarbon date to emerge during the 1950s came from the Niah cave in Sarawak, where in 1958 charcoal found in deposits directly above the skull of a young modern male was dated to just under 40 kyr. This was considerably earlier than any presumed occurrence of modern humans in Europe, which led to doubts of the veracity of the date; these still persist although in the light of what we know today this radiocarbon age is much less improbable than it seemed at first.

The 1950s also witnessed the discovery of a number of significant new human fossils besides those from South Africa already mentioned. Between 1953 and 1957 the archaeologist Ralph Solecki recovered the remains of nine adult and juvenile Neanderthals, associated with a Mousterian industry, from Iraq's Shanidar cave. One of these skeletons was that of an adult male who had suffered, perhaps since birth, from a disabling disease which had deprived him of the full use of his right arm. Without the support of his

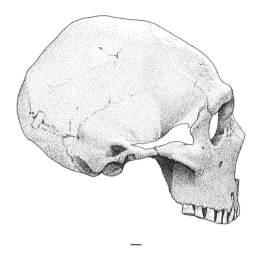
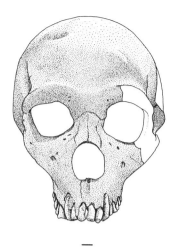

Side and front views of the Shanidar 1 cranium from Iraq. Scales are 1 cm. DM.

social group he could not have survived, but he succeeded in living to an estimated age of forty, an old man by Neanderthal standards. Pollen grains beneath another skeleton suggested that it had been buried with flowers, though this has been disputed. Anatomically and stratigraphically the Shanidar population fell into two groups, specimens from higher in the section having more "classic" features, while those from lower down resembled other members of Clark Howell's eastern group in lacking swept-back cheekbones. Radiocarbon dates made in the late 1950s and early 1960s placed the Shanidar Neanderthals at or beyond the outer limit of the technique; the earlier ones are believed today to date from about 70-80 kyr and the later ones from about 50 kyr.

But quite as important as new Neanderthal finds in the 1950s was the recognition, finally, that the stoop-shouldered, bent-kneed stereotype of these humans created by Marcellin Boule was totally false. Hints to this effect had appeared in the literature over the preceding decades, and in 1955, independently, both the Swiss primatologist Adolph Schultz and the French paleontologist Camille Arambourg stated explicitly that the Neanderthals must have walked fully upright. Schultz noted that the Neanderthal head had been balanced perfectly atop the vertebral column and pointed to the inherent instability of the posture that the Neanderthals were supposed to have adopted. Arambourg looked at various features of the skeleton of the "old man" of La Chapelle-aux-Saints studied by Boule and found no evidence for the latter's deductions. This showed, he declared,

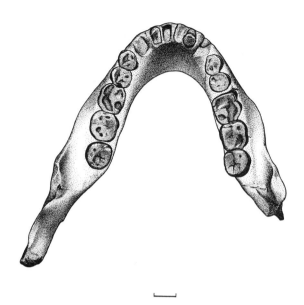

One of the "Atlanthropus" jaws (Ternifine 1) from Tighenif, Algeria. Scale is 1 cm. DM.

that uprightness had to be total to work at all. It's rumored that Arambourg's defense of the Neanderthals was precipitated by his discovery of "Neanderthal-like features" in an X-ray of his own neck; however this may be, Schultz and he were totally vindicated in 1957 when the anatomists W. L. Straus and A. J. E. Cave published a detailed reanalysis of the La Chapelle-aux-Saints skeleton. It turned out that this individual, which had become the standard by which all Neanderthals were judged, in fact showed both osteoarthritic degeneration (which Boule had actually been aware of) and changes due to age. Equally important, many of the differences from modern humans to which Boule had pointed were in fact not differences at all, once the range of variation in modern human populations was taken into account. Indeed, like Arambourg, Straus and Cave could find nothing at all in the skeleton to indicate that its owner had not been a perfectly efficient upright biped. Certain differences there were between Neanderthals and modern people, in the postcranial (body) skeleton, as in the skull; but they certainly did not reflect any lack of a fully upright posture.

Arambourg was involved in a further find of importance made during the 1950s: the discovery in mid-decade of three quite large-toothed lower jaws, some teeth, and a small skull fragment at the site of Tighenif (then Ternifine) in Algeria. In association were found handaxes of Acheulean type and flaked pebbles, as well as a fauna that seemed to indicate an early

. Middle Pleistocene date. Although Arambourg described the human fossils as belonging to the new genus and species *Atlanthropus mauritanicus* (the synthesis and population thinking took an awfully long time to catch on in France), they soon attracted comparison with the mandibles from Zhoukoudian, while many noted differences from the Mauer jaw. It was speculated on this basis that perhaps human evolution in Europe had taken a somewhat different course from that in Africa and Asia; but before opinion had a chance to crystallize on this issue paleoanthropological attention was drawn back once more, and with a bang, to subsaharan Africa.

8

Olduvai Gorge

While the literature of paleoanthropology was already beginning to take its modern shape in the early 1950s, the fossil record of human evolution had to await the end of that decade to follow suit. For it was essentially not until 1959 that Louis Leakey's longstanding interest in Olduvai Gorge began to pay off in the discovery of human fossils there. Following his marriage to the archaeologist Mary Nicol in 1936, the pair devoted themselves to paleontological and archaeological explorations in East Africa; but this work was conducted part-time and on a shoestring until 1948, when the American businessman Charles Boise began to fund their prospections. In 1951 this support allowed work to recommence at Olduvai, where abundant remains were rapidly found of various large mammals. In Leakey's view these were the victims of the early humans who had left behind stone tools in the same deposits. Apart from a few isolated teeth of disputed interpretation, though, there was for years no sign of these humans themselves.

Nobody who has not tried it can fully appreciate the amount of dedication and sheer mental and physical toughness that it takes to continue searching in the broiling sun day after day, let alone year after year, for fossils that nobody can guarantee are there. Generally, you don't "dig" to find fossils, at least not in the first instance; paleontological prospection involves endless prowling or crawling across the landscape, eyes fixed on the ground for the slightest sign of bone at its surface. As fossils erode out of the enclosing sediments they are attacked by the elements and begin to fall to pieces, however complete they may have been to begin with. And complete they will rarely have been, since most fossil bones represent the remains of some ancient carnivore's dinner or have otherwise had a checkered postmortem career. The trick is thus to spot a tooth, or a fractured corner of bone, protruding however slightly among the surface rubble of the eroding deposit. Shadows, stone litter, bumps in the ground, tricks of the light, heat exhaustion, and a host of other things can throw you off. The paleontologist's dream of a landscape covered with fossil bones waiting to be picked up is usually just that, though occasionally it has been realized. And if you are looking for a particular kind of fossil—the remains, say, of the toolmaker of Olduvai rather than of one or another of the vast numbers of large mammals which roamed the same landscape at the same time—the odds lengthen out of sight.

It is thus powerful testimony to the Leakeys' perseverance and optimism that in 1959 they returned yet again to Olduvai, where on a July day Mary Leakey looked once more at a site known as FLK 1. First located twenty-eight years earlier, this site, which had produced an abundance of stone tools, was believed by the Leakeys to represent a "living floor" where the toolmakers had camped and consumed the carcasses of various animals. It lay toward the bottom of Bed I and was thus early in the Olduvai stratigraphic sequence; but despite unhappy experience with his previous claims for ancient *Homo* at Olduvai, Leakey was convinced that the toolmaker of FLK, remote in time though it clearly was, had been a member of our own genus. So convinced of this was he that in 1958 he had identified a large and rather odd-looking isolated molar tooth from higher in the section as that of a truly enormous human child "not of australopithecine type." Broom's associate John Robinson, along with others, later showed that it had belonged to an adult australopithecine. It was with some disappointment, then, that Louis received Mary's news that at FLK 1 she had spotted, eroding from the sediments, a skull whose two visible teeth appeared to be those of a robust *Australopithecus*.

Excavation of the specimen showed first impressions to be correct, however. The skull, almost complete and with a magnificently preserved set of teeth, most closely resembled the robust specimens recovered by Broom from Swartkrans and Kromdraai, but with even more massive chewing (molar and premolar) teeth. In comparison the front teeth (the incisors and canine) were tiny. The braincase was small, with a capacity of about 530 ml, but it bore a large sagittal crest, a ridge of bone along the midline of the skull produced by the attachment of powerful chewing muscles. The face was similarly unprotruding, but much deeper than those of South African *Australopithecus* (or *Paranthropus*) *robustus*. If anything known from outside South Africa was an australopithecine, this was—which created a puzzle, for even if Leakey had been pretty much alone in rejecting an australopithecine ancestry for modern humans, few were prepared to contemplate that any australopithecine had made stone tools: stone tool making,

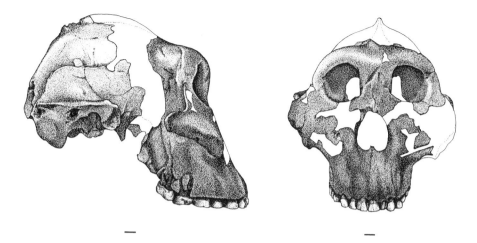

Side and front views of the "Zinjanthropus" cranium (OH 5) from Bed I of Olduvai Gorge, Tanzania. Scales are 1 cm.   DM.

it was thought, was the hallmark of true humanity. Yet the association at FLK 1 seemed to Leakey to be unimpeachable, for the completeness of the hominid specimen contrasted with the broken-up condition of the other vertebrate bones at the site. Evidently the skull, dubbed Olduvai Hominid (OH) 5, had belonged to one of the occupants of the campsite rather than to one of their victims. In the end Leakey compromised between his earlier beliefs and his conclusion that "there is no reason . . . to believe that the skull represents the victim of a cannibalistic feast by some hypothetical more advanced type of man." He described his new specimen as an australopithecine, but of a new genus and species, *Zinjanthropus boisei*, which he claimed to be highly distinct from its South African relatives.

This allocation did not go unchallenged. John Robinson, for one, was quick to claim that the new specimen differed insufficiently from *Paranthropus* to be placed in a separate genus. Robinson, it should be said, was busy at the time defending the notion that a substantial ecological separation between the robust and gracile South African types mandated a generic distinction between them, this in the face of a growing consensus that all should be placed in *Australopithecus*.

Hardly was the ink dry on Robinson's riposte, however, than the plot began to thicken. On a tour to publicize his new find, Leakey had managed to parlay it into support from, among other agencies, the National Geographic Society, then as now a major benefactor of paleoanthropological studies. And with this new backing, work at Olduvai restarted with un-

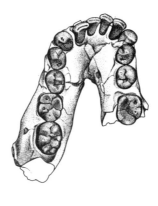

The Olduvai Hominid 7 mandible, type specimen of Homo habilis. *Scale is 1 cm.*   DM.

precedented vigor. At the end of 1959 full-scale excavation commenced at FLK 1 and some adjacent sites, and within a year Leakey was able to announce several more hominid discoveries. FLK 1 yielded a couple of lower leg bones, while at a nearby and slightly earlier locality called FLKNN 1 were found some teeth and skull fragments, some hand bones, and most of a left foot. Whether or not he thought that these fossils represented *Zinjanthropus*, Leakey didn't say; but within weeks of announcing these finds he was back again in the pages of *Nature* with a new specimen from FLKNN 1 that clearly did not.

This was the partial lower jaw of a young hominid with much smaller chewing teeth than *Zinjanthropus* and with relatively larger front teeth: an individual quite reminiscent, in fact, though Leakey was not at pains to emphasize it, of *Australopithecus africanus*. With the jaw were found bits of a braincase which appeared larger than that of OH 5, and Leakey concluded that the whole hominid assemblage from FLKNN 1, skull and postcranial bones alike, belonged to a form "less specialized" than his *Zinjanthropus*. The jaw, braincase, and hand bones later acquired the number OH 7, while the foot became OH 8.

The number OH 9 went to a specimen that Leakey described at the same time but which had been found much higher in the Olduvai sequence, at the top of Bed II. This was a massive long and low skull vault with enormous brow ridges and a brain capacity later estimated at 1,067 ml, right in the middle of the Zhoukoudian range. Leakey didn't commit himself as to the species identity of this "Chellean Man" specimen, so called because it was associated with a stone tool industry formerly known by that name

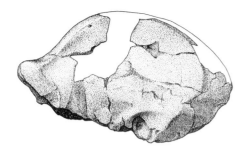
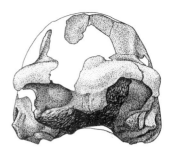

—   —

Side and front views of the Olduvai Hominid 9 skullcap from Bed II of Olduvai Gorge, Tanzania. Scales are 1 cm.   DM.

but which we would now recognize as Acheulean; later authors, however, have been close to unanimous in regarding it as an African *Homo erectus*.

Leakey's noncommittal description of the OH 7 jaw appeared in February 1961. By the middle of that year he was prepared to say a little more. The specimen, he found, showed distinct differences from both *Zinjanthropus* and the South African australopithecines. These boiled down essentially to the length of the premolars, whose chewing surfaces in OH 7 were roundish in outline, rather than oval as in the australopithecines. This was slender

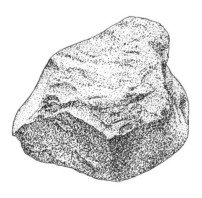

———

View of an "Oldowan" core tool (AMNH 95.3180) in phonolite (a volcanic rock) from Olduvai Gorge, Tanzania. Scale is 1 cm.   DS.

evidence on which to suggest that his "pre-*Zinjanthropus*" was not an australopithecine but instead a "remote and truly primitive ancestor of *Homo*," particularly since the implication of this was to relegate the australopithecines to a side branch of the human evolutionary tree. But this proposal did have the advantage of allowing Leakey to deny once more that the australopithecines were toolmakers. Instead, his new form was the maker of the stone tools of FLKNN 1 and possibly also of the more abundant ones from the *Zinjanthropus* site, FLK 1.

Parenthetically, one might note that these tools were pretty unimpressive things. Termed "Oldowan" after the name Leakey had given the basal stoneworking industry of Olduvai in the 1930s, they consisted for the most part of lava or quartzite cobbles roughly shaped by a few blows with a stone hammer, sometimes while the piece was resting on another stone known as the "anvil." Mary Leakey eventually identified several distinct variants: choppers, discoids, spheroids, and so forth, and viewed all these artifacts as "tools." Nowadays, in contrast, many believe that the flakes struck off these "cores" were more commonly used than the cores themselves as cutting and scraping tools. Higher in the section the basal assemblage gives way to the "Developed Oldowan" in which choppers are fewer and spheroids more numerous. "Protobifaces" appear, which are cores flaked on two sides to produce a cutting edge. At this point more carefully shaped "handaxes" also begin to turn up, pointing toward the Acheulean industry associated with OH 9.

Provocative though they were, Leakey's remarkable claims for his pre-*Zinjanthropus* barely had time to engage the attention of his colleagues when, a week later, he dropped his next bombshell. This concerned the age of the fossils. Leakey himself had vacillated over the dating of Olduvai, but by 1959 he had returned to his original estimate that the Bed I fauna was of early Pleistocene age. No one, of course, knew at that time what the actual duration of the Pleistocene had been, though Leakey did hazard a guess in a popular article that *Zinjanthropus* dated from about 600 kyr ago. Most felt that this figure was not unreasonable. But it was no more than a guess, and it was soon to be outdated by a new technique of radiometric dating: the potassium-argon (K/Ar) method.

Like radiocarbon dating, this new approach depends on the decay of an unstable parent atom to a stable daughter product. But while radiocarbon is the prime example of the "decay" clock, based on the loss of parent atoms, K/Ar is an "accumulation" method, which measures the accretion of daughter atoms. The radioactive potassium isotope potassium-40 (^{40}K) is present as a tiny proportion of all natural potassium. Part of this decays to stable argon-40 (^{40}Ar) atoms, and if a sample of potassium-containing rock is melted in a vacuum these atoms can be counted. The age of the sample can then be derived by comparing this amount to the known isotopic abundance of 40K in natural potassium and applying the decay constant derived from its half-life. Problems may arise, however, because argon gas can get trapped mechanically in some minerals. If such argon is measured along with that derived from the decay of radioactive potassium, the resultant date will obviously be too old. It is for this reason that volcanic rocks are favorites for K/Ar dating, since they crystallize at temperatures at which no mineral can retain any argon. Any ^{40}Ar measured in uncontaminated samples of such rocks must have been derived from radioactive decay, and will give a reliable estimate of how long ago the rock cooled.

Another problem for paleontologists is that fossils themselves obviously cannot be dated in this way: the method simply dates the rocks in which they are found. This is a further reason why volcanic rocks are particularly suitable for paleontological dating, because they are precise stratigraphic indicators: any volcanic event dates from a particular moment in time. Fossils are rarely found in volcanic rocks, and then almost only at the base of ashfalls; but in a continuous sedimentary sequence fossils found in deposits just above or below a particular volcanic layer will be only slightly younger or older than that datable layer—always provided, of course, that the volcanic rock was laid down on the surface of the earth and not intruded, as lavas sometimes are, into layers already buried. Volcanic rocks derived from ashfalls do not present this problem, but even these are from time to time eroded and transported to a new site of deposition subsequent to their initial fall. Usually, however, it is possible to control for the various possible mineralogical and geological pitfalls.

At about 1.3 billion years the half-life of ^{40}K is very long, so ^{40}Ar accumulates rather slowly. This makes the method most suitable for dating very old rocks, though it has on occasion been used to date rocks under 100 kyr old. Beyond the fact that not all fossils are found in geological association with rocks that can be dated by this method, this leaves a gap between the effective ranges of K/Ar and radiocarbon: a gap that has only recently begun to be filled by other methods.

The K/Ar dating method was developed initially in Europe, and it had been used as early as 1950 for dating salt deposits some 20 million years (20 myr) old. Its use on young volcanic rocks came a decade later, however, and Olduvai was the source of some of the first such samples to be dated. Olduvai was a particularly suitable setting for such studies, because the sedimentary deposits there are interlayered with (and indeed in places largely consist of) a series of tuffs (ashfall deposits) and lava flows. The ages of the fossiliferous sediments there are thus bracketed by the dates of the volcanics above and below them. In July 1961 Leakey, together with the Berkeley geologists Jack Evernden and Garniss Curtis, published the first

K/Ar determinations from Olduvai. The results were astonishing. A series of rock samples from tuffs low in Bed I and closely associated with the fossil localities yielded an astonishingly ancient average age of 1.75 myr. This was vastly older than anyone had imagined (though it did have the advantage of alleviating problems of the kind that had worried Leakey back in early 1960 when he contemplated the sheer implausibility of deriving modern humans from a form such as *Zinjanthropus* in what he guessed to be a mere 400 kyr or so). And it did not go uncontested; von Koenigswald, for one, was unhappy with the 2 myr length of the Pleistocene which it implied, and he wrote to *Nature* saying so. But although one of von Koenigswald's co-authors was no less an authority than Wolfgang Gentner, who had produced the first ever K/Ar date over a decade earlier, the Evernden-Curtis date was rapidly accepted and has since been corroborated by many more age determinations from Olduvai and elsewhere.

Studies by the London anatomists John Napier and Michael Day showed that the Olduvai foot was that of a biped; but the hand (since shown to have included a number of nonhuman elements) appeared less like that of later humans. This was uncontroversial stuff, at least at the time; indeed, everyone was treading lightly at that moment because, although there was a lot of skepticism over Leakey's claims, it seemed to most that the FLKNN 1 material was a little slender to argue over. In Wilfrid Le Gros Clark's words, it was placed in a "suspense account." Fortunately, however, material continued to acccumulate from Olduvai. In 1963 a site in the lower middle part of Bed II called MNK produced a partial cranial vault and associated upper and lower jaws, which received the number OH 13. At about the same time a highly fragmentary skull and most of the teeth of a young adult hominid, dubbed OH 16, was recovered from the locality FLK 2 close to the bottom of Bed II. The Leakeys considered both individuals, plus a few other bits and pieces, to belong to the same kind of human as their pre-*Zinjanthropus* specimens from FLKNN I.

What they didn't find at Olduvai was a mandible to match the spectacular *Zinjanthropus* cranium. For this they had to travel some fifty miles northeast from Olduvai, to the west side of Lake Natron. In January 1964, in somewhat younger deposits now known to date from about 1.4 myr ago, they found there an almost complete lower jaw that clearly belonged to what they were now calling *Australopithecus (Zinjanthropus) boisei*. (The parenthetical *Zinjanthropus* indicates a subgenus, and it can effectively be ignored since this category is close to meaningless in paleontology. From now on we'll drop the italics, though the name itself remains a useful convenience for informal reference.)

Emboldened by this more substantial corpus of material Leakey, together with Phillip Tobias and John Napier, finally felt able to broach the question

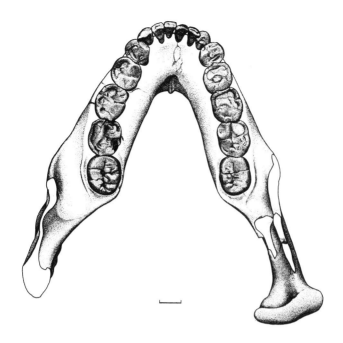

Robust mandible (NMT-W 64-160) from Peninj, Lake Natron, Tanzania. Scale is 1 cm.

DM.

of the identity of the more lightly built human specimens from Beds I and II at Olduvai. In the *Nature* issue of April 4, 1964, they concluded that all belonged to the same species, and that this species was distinct from Zin-janthropus and all other australopithecines (among which, throughout this period, only John Robinson seemed really concerned to make distinctions). Reluctant to place their new form in its own genus, they opted to classify it in the genus *Homo*, as *Homo habilis* ("handy man," reflecting their belief that this was the toolmaker of lower Olduvai). Olduvai Hominid 7—the jaw, braincase fragments and hand bones—became the type specimen, the name-bearer of the species.

The year before, Tobias had reviewed australopithecine cranial capacities and come up with a mean size of just over 500 ml for six *A. africanus* crania; the brain of OH 7, it was estimated on the basis of the very fragmentary material available, had been somewhere around 680 ml in volume. The principal anatomical distinction between *H. habilis* and species of *Australopithecus* was thus a larger brain, and a subsidiary distinction lay in the proportions of the premolars and the anterior teeth. Of course, this move by Leakey and his colleagues involved a radical redefinition of the genus

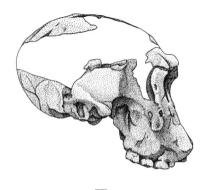
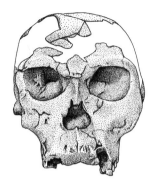

Side and front views of reconstructed cranium OH 24 from Bed I of Olduvai Gorge, Tanzania. Scales are 1 cm. DM.

Homo, particularly since *H. habilis* rather rudely violated Arthur Keith's "cerebral Rubicon" concept. And frankly the revised diagnosis of *Homo* was less than convincing, making reference, for example, to bipedalism (already present in the australopithecines), and leaning heavily on features that were either unknown in the australopithecines or which could be shown to be at the limits of variation in the latter only by lumping robust and gracile together.

In reading their description of *Homo habilis* it is hard to avoid the conclusion that Leakey and his colleagues were swayed principally by the concept of "man the toolmaker," an ancient notion which had been gaining ground since Kenneth Oakley had published the first edition of his booklet of that name in 1949. Leakey in particular was an adherent of the idea that the unique human adaptation was the making of tools, and he was apparently prepared to subordinate anatomical considerations to cultural ones when analyzing the fossil evidence for human evolution. Of course, as we've seen, the term "human" had been deprived of any hope of precise definition as soon as it was realized that people have an evolutionary history. As long as no intermediates were believed to exist between *Homo sapiens* and its closest relatives the apes, there was no problem: people were human and apes weren't, and the keys to humanity lay in the ensemble of those myriad perceived ways in which we see ourselves as different from apes. But as soon as it was acknowledged that modern people had acquired their uniquenesses in the course of a long process, at stages during which some of those differences had been acquired and some hadn't, the question arose (or at least appeared to arise) as to which of them was critical in defining humanity. Keith had fancied the large brain; Leakey was for tool-making; today's Rubicon, as my colleague Mike Rose points out, is bi-pedalism.

An unfortunate spinoff of this mindset was the idea of "hominization": that becoming human was in some way a definable and separate process that could be studied as such. The orthogenetic implications of this notion are clear, and it is in this guise that orthogenesis lingered longest, if in highly attenuated form, in the minds of paleoanthropologists. We'll return later to the precise ways in which this viewpoint conflicts with the actual mechanics of evolution; suffice it to say for the present that cultural associations were a powerful impetus toward the inclusion of *habilis* in *Homo*. And they took on further importance when, at a site called DK 1, low in Bed I at Olduvai, a "rough circle of loosely piled stones" was discovered that the Leakeys interpreted as a possible windbreak constructed by *Homo habilis*. Yet more cultural complexity seemed to provide a yet stronger argument for regarding *habilis* as "human," irrespective of what it actually looked like.

One thing at least was certain, though: that during the early Pleistocene a minimum of two kinds of hominid had been "evolving side by side in the Olduvai region," and perhaps elsewhere, too. Leakey and his coauthors suggested when naming their new species that Broom and Robinson's "Telanthropus" from Swartkrans, by then putatively associated with stone tools and lately transferred by Robinson to *Homo erectus*, might actually belong to *Homo habilis*. Further, there was soon a suggestion that two early humans might have coexisted in southeast Asia as well. Before 1964 was out, Tobias met in Cambridge with von Koenigswald to compare the new Olduvai material with the latter's material from Java. Despite manifest differences in outlook, the two agreed to recognize four successive "grades" in the early and middle Pleistocene of Africa and Asia. The first of these was *Australopithecus* in South and East Africa and just possibly (in the form of highly fragmentary specimens from Java) in Asia. The second grade contained *Homo habilis* from Africa, and maybe also *Meganthropus palaeojavanicus*, a form that had been described by von Koenigswald on the basis of a rather robust fragment of mandible from Java. The third was represented by OH 13 from Bed II, the "Telanthropus" mandible, and a partial mandible and maxilla from Sangiran. The fourth comprised OH 9 and the Tighenif form in Africa, plus the Trinil and Zhoukoudian hominids in Asia. Thus, despite the fact that the Java specimens, in particular, remained hopelessly poorly dated, Tobias and von Koenigswald both saw distinct parallels between human evolution in Asia and Africa from a very early date.

All this, of course, gave Leakey's and Tobias's colleagues plenty to snipe at, especially when Leakey announced to the press that *Homo habilis* stood in the direct line to modern humans while *Homo erectus* represented a mere side branch. The proposal of a whole new species of human, just when the juggernaut of population thinking was rolling so ponderously that almost no amount of variation was seen as too great to be contained within a single species, was generally felt to be a little much to take, even leaving aside Leakey's claims about ancestry. Around Cambridge, where I was just beginning to learn the field at the time, there was much muttering about how there wasn't enough "morphological space" between *Australopithecus africanus* and *Homo erectus* to shoehorn in a new species. And ludicrous as it seems in retrospect—for there is indeed space for a battery of species between the two—this was very much in tune with the spirit of the times. At about the same moment, for instance, the American paleoanthropologist Loring Brace was launching a magnificently sweeping attack on those who would exclude the Neanderthals from the ancestry of modern humans. In an article that commanded wide attention, Brace branded as "anti-evolutionary" virtually anyone who might doubt that hominid history had been an elegantly linear progression from australopithecines through pithecanthropines to Neanderthals and thence to modern humanity. This stately (and almost, by implication, inevitable) process was based on Brace's perception of two consistent and overriding trends in human evolution: increasing brain size (and who could doubt the self-evident advantage of that?) and reduction of the teeth and face due, he felt, to the steady refinement of cutting tools. Brace's brash style attracted a certain degree of opprobrium, but there was widespread sympathy with his point of view, which as we'll see later became raised in certain quarters more or less to an article of faith.

Reactions to Leakey's proposals were thus colored by reluctance to complicate a picture that seemed at last to be simplifying itself. The triumph of the lumping spirit had seen the disappearance into synonymy of most of the names which had been allowed to clutter the human fossil record over the years. Proud of their newfound taxonomic sophistication, paleoanthropologists were naturally hesitant to countenance any reversal of this trend. Reactions were affected too by the common knowledge that Leakey had his own axe to grind, and had indeed been honing it for years. But there were also some more objective reasons to question the conclusions arrived at by Leakey and his coauthors, and overt discussion centered on these. The most important of them stemmed from the fragmentary nature of the Olduvai material. There was no skull of *Homo habilis*; there was not even a complete braincase, or a face. What's more, there was plenty of variation among the fragments known: some had relatively large chewing teeth, for example, while others had smaller ones. Perhaps most important, doubt was expressed about whether the *Homo habilis* type material from Bed I could justifiably be separated from gracile *Australopithecus*. And, with equal reason, it was questioned whether the Bed I and Bed II materials were properly associated in the same species. Criticism did not stop there, of course, but these particular questions can still be legitimately asked thirty years later.

The ritual prescription for alleviating the perplexity of paleoanthropologists is, of course, the discovery of more fossils. After all, it's self-evident that although the known fossil record will never be complete, the more fossils you have the closer to completeness it will be. But the world is never quite that simple, and as usual matters were not helped by the next discovery at Olduvai. This occurred in 1968, when a badly crushed skull (OH 24) was found by P. Nzube at DK East, a site low in Bed I and close to the famous stone circle.

The Leakeys, together with their assistant Ron Clarke, who had painstakingly restored the specimen (and was to become famous for many similar feats), found predictably enough that it belonged to the genus *Homo* as enlarged in 1964 with the addition of *habilis*. But where John Robinson had concluded that the Bed I and Bed II fossils lay in the same lineage but in different species (of *Homo: H. transvaalensis* and *H. erectus*), the Leakeys still saw no reason to suppose that two different species of *Homo* had been present at Olduvai during Bed I and lower Bed II times. Moreover, in 1965 the paleoanthropologists Elwyn Simons and David Pilbeam, both then at Yale, had reviewed the Olduvai material and concluded that between Bed I and Bed II times dental reduction had taken place in the gracile lineage while brain size had remained rather constant. Their provisional conclusion was that the Bed I material was probably not distinct from *Australopithecus africanus*, while the Bed II specimens belonged to the same species as Robinson's *Telanthropus capensis*, potentially a member of *Homo*. But they were also prepared to entertain a variety of other possibilities, which only goes to show how much difficulty objective observers were having in making sense of the rather scanty fossil record available during the 1960s. This record was sufficient to show that things might well have been more complicated than the nomenclatural spring-cleaning in the wake of the Synthesis suggested; it was, however, inadequate to support any clear-cut new picture.

It's hardly surprising, then, that it took some fifteen years and the discovery of a variety of new fossils for paleoanthropologists to become at all comfortable with the idea of *Homo habilis*. And it took longer still to see how recognizing this new species would pose as many problems as it appeared to solve. Nonetheless, it is possible in retrospect to see that, largely through the efforts of the Leakeys, the human fossil record had begun, by the mid-1960s, to take on the outline that is familiar today.

9

Rama's Ape Meets
the Mighty Molecule

While most members of the paleoanthropological establishment were industriously debating the significance of the Leakeys' newfound Olduvai fossils, Elwyn Simons of Yale was busy looking further back in time, toward the origin of the hominid lineage. When Simons arrived at Yale's Peabody Museum of Natural History in 1960 he inherited the curatorship of a collection of fossils made some thirty years earlier in the Siwalik Hills of northern India by G. Edward Lewis, then a Yale graduate student. Among these fossils were two jaw fragments that Lewis described in 1934 as "a much closer approximation to the Hominidae than other genera hitherto recorded", even though they were of late Miocene age (around 7-8 myr old as we now know, though they were believed in the 1960s to be about 12 myr old). He made one of them the type specimen of a new genus and species, *Ramapithecus brevirostris* (Rama's short-faced ape), named for the Hindu god. However, it was only in his doctoral dissertation (submitted in 1937) that Lewis went so far as to claim that his *Ramapithecus* partial upper jaw actually belonged to the human family—a claim he clung to in spite of a ringing denunciation in 1935 by the Smithsonian's Hrdlička, who wrote that *Ramapithecus* could not be "established as a hominid, that is, a form within the direct human ancestry." With unintended irony, though, Hrdlička also declared that it was "nearer to man than . . . *Australopithecus*."

At least in part because his dissertation remained unpublished, Lewis's assertion languished in obscurity until 1961, when Simons ran across the *Ramapithecus* type specimen in the Peabody collections. After some thought he concluded that it was indeed hominid, and he promptly described it in print as "a forerunner of Pleistocene Hominidae." A year later Louis Leakey gave the name *Kenyapithecus wickeri* to a pair of maxilla fragments from 14 myr old deposits at Fort Ternan in Kenya. At first he simply announced that his new species "exhibits a marked tendency in the direction of Hominidae", but he soon came to regard it unhesitatingly as a human ancestor. Simons agreed, and indeed in 1963 he declared (to Leakey's chagrin) that the Kenyan and Indian specimens belonged to the single species *Ramapithecus brevirostris*. Subsequently a variety of other teeth and bits and pieces of jaw, lower as well as upper, and from Europe and China as well as from India and Kenya, also became incorporated into *Ramapithecus*.

Given the sparse material available in the early 1960s, the argument that *Ramapithecus* was a human precursor had to rest on the shape of the teeth and palate. As in *Australopithecus* the molar teeth appeared squarish, with low crowns and rather flat chewing surfaces, while the canines and the incisor teeth (known at that time only from their empty sockets) appeared to have been much smaller than those of living apes. Moreover, the chewing teeth turned out to have quite thick enamel coatings, something not found in the African apes. No complete palate was preserved, but as reconstructed by Simons the toothrows of *Ramapithecus* resembled those of modern humans in describing a parabolic curve. In contrast, the toothrows of apes form a U-shape. The divergence of its toothrows made the palate of *Ramapithecus* appear relatively short from back to front, which in turn suggested, as Lewis had, that this hominoid had possessed a humanlike short face rather than the protruding muzzle of a modern ape.

As long as only a few suggestive fragments were known this reconstruction was quite plausible as far as it went. But after David Pilbeam joined him as his graduate student Simons rapidly went well beyond his early cautious statements about the ancestral human status of *Ramapithecus*. Reduction of the front teeth, wrote Pilbeam and Simons jointly in 1965, implied tool use "because smaller front teeth require the use of other means to prepare food, either animal or vegetable." What's more, "The evolutionary shift in a major adaptive zone indicated in the case of *Ramapithecus* by its reduced snout and anterior teeth (premolars, canines and incisors) . . . may correlate with the incipient development of bipedality." In sum, "the commitment to a hominid way of life had been made by the late Miocene, and our earliest known probable ancestors . . . might already have adopted a way of life distinct from that of their ape contemporaries."

This was a great deal to infer from a handful of jaw fragments, but the notion that a small-brained but upright and tool-using human precursor had emerged from the pack of hominoids rather far back in time proved a seductive one indeed. For, albeit on the slender basis of a couple of features of the dentition, Pilbeam and Simons had built up a portrait of a putative human ancestor that closely reflected prevailing suppositions about how the human lineage had emerged. To take one example, an energetic debate went on more or less throughout the 1960s over the significance of canine tooth reduction in humans. Most paleoanthropologists of the period supported Darwin's original idea, published in his *Descent of Man*, that human canines had become reduced when tools supplanted them in fighting and display. Further, they tended to believe, again with Darwin, that tools could not have been made or used if the arms had been preoccupied with locomotion. Those who demurred from this orthodoxy generally affirmed the importance of toolmaking but considered that this behavior should be seen simply as part of a wider cultural context; for by this time the notion of humans as essentially cultural beings had gained broad ascendency. In this intellectual climate, it was not surprising that by the end of the 1960s it had become widely accepted among paleoanthropologists that both the genealogical and the behavioral roots of the human lineage lay back as far as 15 myr ago—and further, if you believed Louis Leakey, who by 1967 was pushing a 20 myr old alleged species of his *Kenyapithecus* as a human ancestor.

As more material of *Ramapithecus* was discovered, however, this edifice proved to be top-heavy. In 1973 the paleoanthropologists Alan Walker and Peter Andrews, then both based in Nairobi, published reconstructions of the upper and lower jaws of *Ramapithecus wickeri* from Fort Ternan. New material (actually, a partial lower jaw from Fort Ternan previously identified as an ape) allowed the shape of the dental arcade to be determined with some confidence for the first time—and it proved to be of a shallow V-shape rather than the parabolic curve of Simons's reconstruction. This in itself wasn't a problem, for there were actually plenty of specimens of *Australopithecus* available to show that modern-style parabolic dental arcades weren't characteristic of those by now undoubted hominids, either. But it was symptomatic of a shift in the climate of opinion. The year before a couple of suggestions had already been published that the giant hominoid *Gigantopithecus* (another form with large, flat chewing teeth and reduced incisors and canine) might make a better candidate than *Ramapithecus* for hominid ancestry. And at around the same time some paleoanthropologists began to cast doubts more directly on the claims of *Ramapithecus* to be hominid. In 1973, for example, Christian Vogel of Germany's University of Göttingen stated flatly at a conference in Chicago that "all hitherto discussed features [of Lewis's type maxilla] . . . are not sufficient to warrant the inclusion of *Ramapithecus* in the Hominidae," a refrain which was later taken up by Milford Wolpoff and some of his graduate students at the University of Michigan at Ann Arbor.

Eventually, under fresh scrutiny and with new fossils available for comparison, the alleged uniquely humanlike traits of *Ramapithecus* were seen to fall away one by one, until almost the only feature that united this genus with *Australopithecus* and *Homo* was the great thickness of the enamel on the molar teeth. Then, however, it was found that the undoubted Miocene ape *Sivapithecus*, which had been viewed by Pilbeam and Simons themselves as a potential ancestor of the orangutan, also shared this unusual feature. So did the orangutan, as well as *Gigantopithecus* and various other fossil apes. Finally, in 1980, Peter Andrews and his Turkish colleague Ibrahim Tekkaya delivered the paleontological coup de grâce. They showed on the basis of the accumulated new fossils that the teeth of *Ramapithecus* and the ape *Sivapithecus* in fact resembled each other so closely that placing them into separate genera was not reasonable—as Leonard Greenfield of Temple University had been urging for some time. If one of these genera was an ape, so was the other. And they had a new fossil, a partial face of *Sivapithecus* from Sinap, in Turkey, that showed that this beast was remarkably orangutanlike. The conclusion was clear: *Ramapithecus* was just another ape, a member of a group related to the orangutan.

By this time, Pilbeam himself had already jumped off the *Ramapithecus*-as-hominid bandwagon. On an expedition to Pakistan in 1976 a member of his field crew had found a lower jaw of *Ramapithecus* that conserved the shape of an entirely nonhumanlike dental arcade. This precipitated an influential rethinking on his part, a process that was completed during his 1979-1980 field season in Pakistan with the discovery of a more or less complete 8 myr old face and palate of *Sivapithecus*. This remarkable specimen dramatically confirmed the uncanny resemblance between the dental and facial features of *Sivapithecus* (in which genus he, too, came to include *Ramapithecus*) and those of the orangutan. Pilbeam's dramatic about-face on the matter of *Ramapithecus* as a prehuman is probably the most famous and reputation-boosting case of its kind in paleoanthropology since Emile Cartailhac's legendary "Mea culpa d'un sceptique" of 1902, when the French savant publicly renounced his opposition to the authenticity of the Altamira paintings. In any event, by the early 1980s *Ramapithecus* was no longer a contender for status as an early human, leaving *Australopithecus* with no obvious antecedent in the fossil record.

Pilbeam's change of heart was not, however, brought about solely by the accumulating evidence of the fossil record, or even by the arrival of more sophisticated ways (which we will discuss in the next chapter) of analyzing that record. For during the 1970s an entirely new approach had been steadily penetrating into the traditionally fossil-based domain of paleoanthropology. This was molecular systematics, which actually had a respectably long pedigree, tracing its ancestry to work performed by the Cambridge University bacteriologist George Nuttall in the early years of this century. The blood constituents of different species should, Nuttall reasoned, pretty much resemble each other in proportion to the closeness of the genetic—

hence phylogenetic—relationships between those species. What's more, it should be possible to determine the closeness of those relationships by the strength of their immune systems' reaction to each other's blood proteins.

The vehicle chosen by Nuttall to investigate this idea was the "precipitin test," based on the fact that an individual's immune system will produce antibodies to fight the blood proteins of another individual as if they were invading agents of disease. If you take a sample of blood from a member of species A and inject it into a member of closely related species B, B will produce antibodies to fight the blood proteins of A. If you now take the blood serum of B (at this point known as the "antiserum" since it is full of antibodies to A) and inject it into a variety of other animals, you can see how strongly the blood of these other species reacts. The stronger the reaction, the cloudier (with precipitin) the mixture of bloods will become. And, because closely related species share more sites on their blood molecules that can be attacked by antibodies produced to their close relative, the resulting cloudiness will increase in proportion to the similarity of the blood samples being tested. Nuttall eventually tested a very large variety of animals using these methods; he found among other things that a closer "blood relationship" existed between humans and the large-bodied apes than between humans and the Old World monkeys, and that the New World monkeys were yet more distantly related to us. This was hardly a momentous revelation—indeed, Nuttall emphasized that this had been Darwin's own view—and there the matter rested for more than half a century, even as information continued to accumulate on the mechanisms of heredity and, ultimately, on its molecular basis.

In the early 1960s Morris Goodman of Wayne State University began to resuscitate the idea of comparing molecules, via immunological reactions, as a means of unraveling evolutionary relationships. And what he found, using techniques a lot more sophisticated than any available to Nuttall, went completely against received wisdom in anthropology. While every conventional classification of the higher primates placed humans in one family—Hominidae—and the great apes together in another—Pongidae—Goodman concluded that the African apes were so similar immunologically to humans that all should belong to the single family Hominidae. Further, while the African apes showed close immunological affinities to each other and to humans, the Asian apes, the orangutan and the gibbon, diverged from this group. Goodman then went on to bolster this argument with studies of various blood proteins using a process called electrophoresis, which sorts molecules by their size and weight. Again, he found that the African apes clustered with humans rather than with the Asian apes, supporting the idea that chimpanzees and gorillas should be clasified in Hominidae along with humans, leaving the orangutan to occupy the family

Pongidae in lonely state (the gibbons had long been given a family of their own). However, Goodman also found that while some molecules were most similar in humans and chimpanzees, other human blood proteins most closely resembled those of gorillas—thereby launching a debate about the exact relationships among the three that continues to this day.

Back in the early 1960s, however, Goodman's problem was getting anyone to listen to him at all. His findings went so flatly against received wisdom that traditional primate systematists rejected his findings out of hand. As an innately cautious scientist, Goodman was greatly concerned to understand just why it was that his conclusions about human-ape relationships differed so clearly from those of the morphologists (which was, in fact, partly because of inherent difficulties due to the close relationships among all the species concerned and, more important, because a truly rational procedure for the analysis of morphology had yet to come on the scene—of which more later); and he always sought to reconcile the two lines of evidence, morphological and biochemical.

Not so Goodman's fellow biochemists Vincent Sarich and Allan Wilson, of the University of California at Berkeley. During 1966 and 1967 these researchers published a series of studies that employed a new and powerful immunological technique to compare the important blood protein known as albumin among a variety of primate species. This method, known as micro-complement fixation, provided them with a quantitative measure of albumin similarity, from which they were able to construct a scheme of relationships among the species concerned.

In general, there was broad agreement between the relationships thus established and those that had been proposed by Goodman. What really upset the morphologists, however, was that Sarich and Wilson used the internal consistencies that they perceived in their data set to argue that the albumin molecule changed at a constant rate. Thus, they claimed, the "immunological distances" between species based on this molecule could be used to calculate the time that had lapsed since they had shared a common ancestor. This added insult to injury. Not only were the biochemists usurping the traditional function of the morphologists in determining relationships, but they were now moving in on the paleontologists, the guardians of the time element in evolution!

Still worse, at a time when the 14 myr old *Ramapithecus* had finally established ascendency as a human ancestor, Sarich and Wilson's time estimates were totally at variance with this prevailing wisdom. The "molecular clock" needs to be calibrated using one date taken from the fossil record, and the date chosen by Sarich and Wilson for this purpose was one that was palatable to most paleontologists of the day: 30 myr for the last common ancestor of the apes and the Old World monkeys. The date that this yielded for the last common ancestor of humans and the African apes was not, however, palatable at all: about 5 myr. In later publications Sarich and Wilson softened this divergence date a little, but they yielded nothing on principle. Indeed, in 1971 Sarich wrote, in one of the most breathtakingly provocative and undiplomatic statements in the history of human evolutionary studies, that "one no longer has the option of considering a fossil older than about eight million years as a hominid *no matter what it looks like.*"

As biochemical techniques proliferated it became clear that different methods and different molecules did not produce monolithic results; here was no magic silver bullet for systematists. Indeed, to this very day there is among the primates a whole variety of molecular phylogenies on offer. Most of these differ principally in detail, though in some cases there is profound disagreement; and molecular systematists argue among themselves at least as much as morphologists do. Nonetheless, it might seem surprising that back in the infancy of the debate, and in the face of united morphological opposition to molecular systematics, it was Morris Goodman who was perhaps Sarich and Wilson's most vociferous opponent. As ever trying to find some common ground between the molecular systematists and the morphologists, Goodman suggested that an extended time scale for the human family might indeed be compatible with the biochemical data. Such a reconciliation was possible if Sarich and Wilson were wrong about the constancy of molecular change, and if this had in fact slowed down among the hominoids compared to primates in general. The precise argument for such slowdown depended on the fact that hominoids have a long gestation period combined with a particularly intimate contact between the fetal and maternal bloodstreams during that period. Any immunological incompatibility between the mother and the fetus over such a long period would be deeply injurious to the latter; thus, Goodman suggested, natural selection would have acted to reduce such possibilities, and hence to reduce the rate of immunological change.

But as debate between the biochemists raged, some morphologists, Pilbeam and Andrews among them, were beginning by the early 1980s to see merit in Sarich and Wilson's short phylogeny; and there's little doubt that it wasn't just new fossil evidence but also the new biochemical perspective that had speeded their reappraisal of the significance of *Ramapithecus* in human evolution. Of course, biochemistry is handicapped, even more than is the comparative anatomy of living species, in being unable to fill out the story of human evolution with the names, or the appearances, or the environments and adaptations of our various ancestors: only the fossil record can do that. Moreover, there's no prospect that useful molecular data will be extractable from higher primate fossils of any ancientness in the near future. But from today's viewpoint Sarich and Wilson's divergence times look pretty accurate and, more important, molecular systematics can indeed help to sort out the basic relationships among the living species that make up the group of animals to which we belong. It was certainly the use of molecular methods that led initially to the questioning of the great apes as a monolithic group distinct from ourselves, and hence gave rise to the big nomenclatural question with which we are now wrestling. Let's digress for a moment and look at this question.

Knowing what to call ourselves used to be simple: together with *Australopithecus* et al. we were hominids, belonging to the family Hominidae; the great apes formed the separate family Pongidae, and the gibbons another called Hylobatidae; all three families together comprised the superfamily Hominoidea. Within Hominidae there were two subfamilies: Australopithecinae for the earlier forms, and a rather superfluous Homininae for members of the genus *Homo*. One sighs today for such simplicity. Now, even if one accepts that the African apes are our closest living relatives (which not everybody does; Jeffrey Schwartz of the University of Pittsburgh is a particularly energetic advocate of the idea that the orangutan occupies that position), it is by no means clear that the two African apes form a subgroup of Hominidae distinct from ourselves. Neither the molecular nor the morphological approach has yet been able to sort out definitively how to resolve the human-chimpanzee-gorilla split; any combination of closest relatives still seems possible. This is particularly inconvenient because it leaves us in doubt whether to classify us and our fossil relatives as a distinct group at the level of the tribe Hominini or of the subtribe Hominina. And it leaves us with an equal uncertainty about how to refer to what used to be the subfamily Australopithecinae. For the rest of this book we'll continue calling them australopithecines for the sake of convenience, while realizing that this isn't any longer quite kosher. We'll also continue using the familiar term "hominid" for the australopithecines plus members of *Homo*.

Of course, it's unfair to blame molecular systematics for this dilemma, which simply results from a more accurate perception of humanity's place in nature—or, to be more precise, of the difficulties inherent in determining that place with total precision. There's no doubt in my mind that the problem would in any case soon have become apparent as a result of advances in the analysis of morphology which began to filter through into paleoanthropology in the 1970s. We'll look at those advances in another chapter; suffice it to say for the present that the fragmentary *Ramapithecus* could never have achieved distinction as humanity's remote ancestor had we known in the 1960s what we know today about how to extract information from fossils.

10

Omo and Turkana

The power of a persuasive paradigm should never be underestimated, and during the 1960s the concepts of the New Evolutionary Synthesis combined with notions of culture as the basic human attribute to produce a new perspective—even dogma—on the human evolutionary process as a whole. Before that time, of course, various suggestions had been made for reducing the number of named species of early humans; but this had usually been done on a case-by-case basis. It was only during the 1960s that the notion became popular that there *could* only ever have been one species of hominid extant at any one point in time. In formulating this concept, the advocates of the "single species hypothesis" trawled widely for their theoretical underpinnings. From ecology they borrowed the notion that every species was defined by its ecological niche, and that "competitive exclusion" would ensure that no two species could for long occupy the same niche. From the New Synthesis came the idea that evolution consisted exclusively of gradual change from generation to generation. And from anthropology they took the concept that humanity was defined by the possession of culture rather than by any particular physical attribute.

Loring Brace was instrumental in getting this intellectual bandwagon rolling. Generalizing to paleoanthropology as a whole the accusations of "hominid catastrophism" that he had already leveled against those who would deny ancestral human status to the Neanderthals, Brace was arguing as early as 1965 that the australopithecines (not Robinson's Telanthropus) were the makers of the Swartkrans stone tools. This, he said, meant that they were culture-bearing, and thus by definition belonged in *Homo*. What's more, culture in itself constituted an ecological niche, and because of the competitive exclusion principle, no two culture-bearing hominids could exist at the same time. The entire story of human evolution thus boiled down to a simple succession of four stages (his Australopithecine, Pithecanthropine, Neanderthal, and Modern), that were only arbitrarily defined by breaks in the known fossil record. Specifically, he claimed that they "formed points in what was in fact a continuum." There is indeed a certain beauty in simplicity, and here was a seductively simple idea that combined a variety of fashionable elements. Nonetheless, Brace's voice went largely unheeded until Milford Wolpoff, probably the loudest voice in the business, took up the single-species cudgels.

In the years following 1967 Wolpoff proclaimed the single-species message in paper after paper and at meeting after meeting, and rapidly spawned a generation of like-minded intellectual offspring. Wolpoff deployed and elaborated all the arguments Brace had made, and quite a few more. Among other things, he demonstrated to his own satisfaction that there was so much morphological overlap among the various samples of South African australopithecines that they probably all belonged to the same species. There are in fact other reasons for this apparent overlap; but in the late 1960s not a few paleoanthropologists were prepared at least to entertain the possibility that the differences between the robust and gracile australopithecines were due to sexual dimorphism: size and shape differences between the sexes (which in the bones of modern humans, if not apes, are relatively minor). And while others mumbled about how remarkable it was that all of the females had died at once at Sterkfontein, while all the males had waited around another half-million years before stampeding across the valley to become extinct at Swartkrans, it was clear that clinching evidence one way or the other was going to have to come from an expanded fossil record. Fortunately, there was not long to wait.

In 1966 the Ethiopian Emperor Haile Selassie made a state visit to Kenya. While there he met Louis Leakey, who showed him some of his fossil discoveries from Olduvai. When the emperor inquired why there were no such things known from his own country, Leakey declared that they were undoubtedly there; it was only necessary to look for them. An invitation to do so was rapidly forthcoming. This was not in fact the first time that Leakey had contemplated fossil-hunting in Ethiopia; in 1959 he had helped Clark Howell with a survey trip to the Plio-Pleistocene deposits of the Omo basin, just to the north of Ethiopia's border with Kenya, and now Leakey called upon Howell again. And since Camille Arambourg had visited the Omo some decades earlier, Leakey also invited the French paleontologist to participate in the Omo endeavor. For reasons of age and health, however, neither Leakey nor Arambourg was heavily involved with the expedition; instead, Leakey was represented by the Leakeys' son Richard and Arambourg by a young French paleoanthropologist, Yves Coppens. Richard Leakey, it should be said, had been overdosed on paleontology as a child, and was at that time making his own career as a white hunter; his father called upon him principally because of his demonstrated skill at organizing expeditions to remote places.

During the first field season, in 1967, Richard rapidly came to see that he was very much a junior partner in the enterprise, a fact that became highly evident when he was assigned the least interesting area of deposits to survey (although he did manage to recover a pair of partial human skulls, from deposits probably around 125 kyr old, which later figured importantly in discussions of the origins of anatomically modern humans). Frustrated, he borrowed a helicopter that had been hired by the expedition, and flew south into Kenya, over the eastern shore of Lake Rudolf (since renamed Lake Turkana). Landing, he found fossils eroding out abundantly from the sandstone: and here, he decided, he would make an independent name for himself. We'll look at the historic results of this decision in a moment.

The French-American expedition continued work in the Omo until 1974, when political conditions in Ethiopia rendered further work difficult. Although this research did not result in a very bountiful harvest of hominid fossils, it was extremely important for a couple of reasons. First, abandoning the venerable Leakey-style image of the lone paleontologist, Howell insisted from the start on assembling a team of specialists—geologists, geochronologists, human paleontologists, paleontologists of various other specialties, archaeologists, and so forth—to do the work necessary to understand the complex geology of the area and to recover and analyze the fossils recovered from it. Howell's multidisciplinary approach served as a model for all later enterprises of this kind.

Second, the Omo deposits (which originally were thought to represent a short time-span just older than Olduvai) turned out in fact to provide a sort of yardstick of Plio-Pleistocene history. They consist of mostly riverlain sediments of great thickness that chart the geological history of the area over almost four million years. Since these layers have been tilted by earth movements (like a layer-cake tipped on its side), modern fossil hunters moving across the Omo landscape are literally moving across time. Interspersed with the fossil-containing river sediments are numerous lava flows and tuffs (layers of volcanic ashfall) that are ideal for chronometric dating. Frequent faulting made the geology extremely difficult to interpret, but it also ensured that rocks from throughout the sequence were exposed at the surface and thus collectable by paleontologists. Throughout the thick section of rock thus exposed, abundant fossil faunas were found that bore witness to the changes in animal life in the region over a long period. Howell's insistence upon exact documentation of where each fossil had come from in the section, combined with the frequency of datable tuffs,

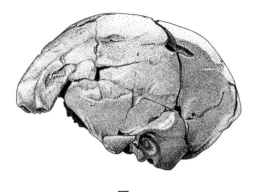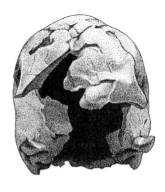

Side and front views of the Omo 2 braincase from the Kibish Formation, Omo Basin, Ethiopia. The more "archaic" of the two Kibish specimens. Scales are 1 cm.   DM.

made calibration of this sequence of biological change particularly precise. Especially useful were the fossil pigs, studied in detail by the paleontologist Basil Cooke. The various combinations of extinct pig species were so accurately dateable and frequently replaced that they could be used for dating sediments elsewhere when chronometric dates were lacking or in doubt.

Quite a few human fossils were also found in the Omo, though most of them consisted of isolated teeth because the bulk of the deposits were laid down by relatively fast-moving waters, in which prospective fossils tend to get broken or destroyed. Following the last field season, Howell analyzed these specimens and found four kinds of hominid represented. The most abundant of these was a massive robust hominid that most closely resembled Leakey's Zinjanthropus from Olduvai, and Howell thus placed it in the species *Australopithecus boisei*. Such fossils dated to between about 2 and 1 myr ago. The oldest teeth turned up between about 3 and 2 myr ago and were said to resemble *Australopithecus africanus* from South Africa. Later teeth (about 1.85 myr) were attributed to *Homo habilis* (helping to allay Howell's doubts about this species), and later ones yet to *Homo erectus* (about 1.1 myr). In addition to the human fossils, the team's archaeologists also found various occurrences of crude stone artifacts in deposits dating from more than 2 myr ago.

The hominid fossils of Omo were, however, overshadowed by the finds Richard Leakey was making along the eastern shore of Lake Turkana. Following his retreat from Ethiopia, Richard had returned to east Turkana in 1968 with a small team of specialists. This survey confirmed the great fossil

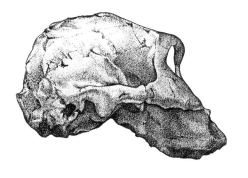
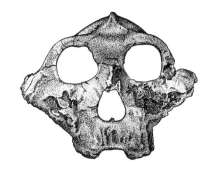

—   —

Side and front views of the cranium KNM-ER 406, from the KBS Member, Koobi Fora, Kenya. Scale is 1 cm.   DM.

potential of the area, though few fossils were actually collected at that time pending better knowledge of the local geology. This was actually to prove a little ironic, since it eventually turned out that fossil locality records during the first few years of Turkana collecting were more than somewhat deficient—but we'll pick up on that story later. In any event, it was in 1969 that the younger Leakey, by then Administrative Director of the National Museum of Kenya, really hit paydirt. That year two hominid skulls were found, one badly crushed, the other almost complete except for the teeth. The better preserved fossil, known as KNM-ER 406 (an abbreviation of its museum identification: Kenya National Museum, East Rudolf, specimen no. 406), was remarkably like the Zinjanthropus cranium, except that its face was much shallower from top to bottom, confirming the suspicion of many that OH 5 had been exceptional in this feature. It was thus allocated to the species *Australopithecus boisei*. The other cranium, ER 407, was more lightly built and was reported by Leakey as most probably belonging to a species of the genus *Homo*. Another highlight of the 1969 season was the discovery of crude stone tools within a datable tuff (the so-called KBS tuff), and Leakey felt it likely that ER 407 represented the toolmaker. The tuff itself had yielded a potassium-argon age of 2.6 myr, so already east Turkana appeared to be yielding evidence of separate *Homo* and *Australopithecus* lineages at a remarkably early date, a conclusion guaranteed to warm Louis Leakey's heart. The scene was, however, already being set for controversy: some years later ER 407 was reidentified as a female *A. boisei*. Eventually, both skulls turned out to be younger than originally thought, having come from sites that lay stratigraphically above the KBS tuff, not from below it

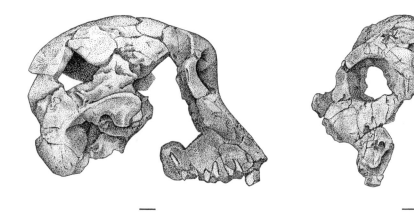

Side and front views of the cranium KNM-ER 732, from the KBS Member, Koobi Fora, Kenya. Scales are 1 cm.   DM.

as initially reported; and the 2.6 myr date for that tuff itself sparked a long-running argument that was eventually resolved by a reassignment to 1.9 myr.

By 1970 the younger Leakey had assembled an impressively diversified team to work on what was now called the Koobi Fora Research Project, from the name of the site where the main camp had been established. And as the fossils began to roll in as if on a conveyor belt, each new find appeared to support Richard's initial conclusion that *Homo* and *Australopithecus* lineages had been separate at a very early date. The jewel of the 1970 season was a partial cranium (ER 732) of what everybody was to come to view as a female *A. boisei*; it was less massive than the male ER 406, and lacked the cranial cresting, yet it had much the same overall look to it. Certainly, ER 406 and ER 732 did not contrast with each other in the same way as did, say, *Australopithecus africanus* and *A. robustus* of South Africa. There were also several very large lower jaws, plus some that were much smaller but did not plausibly belong in the same species as the ER 732 cranium. Here at last was more or less incontrovertible evidence that robust and gracile early hominids were not simply males and females of the same species. Yet resistance remained from two quarters: from advocates of the single-species hypothesis, and from those who were unhappy with the idea of *Homo habilis*. At times the two became confused; for example, straight-forward opponents of *Homo habilis* occasionally trotted out the theoretical arguments for the single-species idea in support of their position.

The 1971 fossil haul from Koobi Fora included some specimens that were

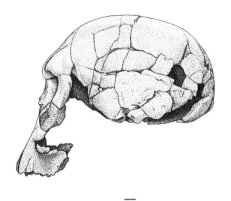
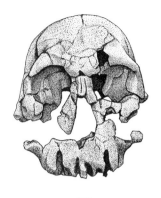

Side and front views of the cranium KNM-ER 1470, from the Burgi Member, Koobi Fora,
Kenya. Scales are 1 cm.   DM.

later to assume an unexpected importance. These included the lower jaw
dubbed ER 992, which came from later deposits than the fossils we've discussed so far. It also looked remarkably like *Homo erectus*, though with well-
established caution Leakey referred to it simply as *Homo* sp. (unspecified
species of *Homo*). Radical rethinking thus had to await the 1972 field season
at Koobi Fora, which produced the now-famous ER 1470 skull. This was
found in hundreds of fragments, but painstaking reconstruction by Alan
Walker and Richard's wife Meave revealed that much of the cranium had
been preserved, although contact between the braincase and what remained
of the face was minimal, and the teeth were entirely gone. What was left,
however, was entirely distinctive. The braincase was unexpectedly large; it
had, Leakey reported, contained a brain of at least 800 ml in volume (later
scaled down to about 750 ml). This contrasted with considerably under 550
ml for even the largest australopithecines, and an estimated 640 ml for
Louis Leakey's original *Homo habilis*. On the other hand, the face was pretty
flat, and the palate was blunt and wide in a way that Leakey found reminiscent of *Australopithecus*; and the chewing teeth, to judge by what was left
of their roots, had been pretty large. What's more, the appearance of the
specimen changed dramatically according to the angle at which the face
was joined to the rest of the cranium.

All this left plenty of room for interpretation, and Alan Walker, for one,
thought the specimen might as readily be assigned to *Australopithecus* as to
Homo. In the end, however, the brain carried the day. Leakey assigned the
specimen to the genus *Homo*, but to an undetermined species. To Leakey's
mind it wasn't the same thing as *Homo habilis* from Olduvai, for two reasons

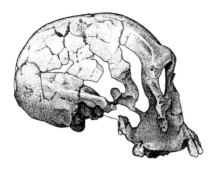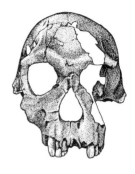

Side and front views of the cranium KNM-ER 1813, from the Burgi Member, Koobi Fora, Kenya. Scales are 1 cm.   DM.

he was prepared to admit to. Not only was the brain of ER 1470 considerably larger than that of Olduvai *Homo habilis*, but the Turkana fossil was also thought to be considerably older. From its stratigraphic position below the KBS tuff, its age was estimated at 2.6-2.9 myr. A supplementary reason for the younger Leakey's reserve may have been a natural reluctance to plunge into the acrimonious debate that still surrounded *Homo habilis*. If this was so it's particularly ironic that ER 1470 was, rightly or wrongly, ultimately responsible for the general acknowledgment of *Homo habilis* as a valid species of early human. What was pretty clear from the start was that 1470 didn't fit at all comfortably into any of the generally accepted early human species; but it certainly could not be ignored, and the sheer lack of definition of *Homo habilis* made it a useful slot into which this remarkable specimen could be squeezed. Suddenly, the new species had a purpose.

Meanwhile, however, new discoveries at East Turkana were muddying the waters even further. The 1973 field season turned up two more crania: ER 1805 and 1813. The first of these was both quite fragmentary and found encased in a tough matrix from which it took some time to emerge fully. Since its discovery it has been interpreted as belonging to almost every species even remotely possible, the latest analysis suggesting that its affinities lie with Olduvai *Homo habilis*. ER 1813 is a broken but quite complete cranium, with many of its teeth intact. Quite lightly built, and with a brain volume of not much more than 500 ml, it nonetheless has teeth reminiscent of some *Homo habilis* from Olduvai. In describing it in 1974 Leakey was impressed by its resemblances to South African *Australopithecus africanus*, but others have since pointed to differences from the latter, particularly in the face. Back in 1973 most paleoanthropologists found it hard to conceive that a member of *Homo* could have had such a small brain, but eventually Clark Howell opined that 1813 might be a female *Homo habilis*, an assignment toward which most subsequent opinion has also inclined. This assignment more than anything else reflects the usefulness of having around a basket called *Homo habilis* into which paleoanthropologists could sweep a lot of fossil loose ends. And of course, the more this basket swelled, the less biological meaning it possessed. Richard Leakey himself has, indeed, never been comfortable with it, and has recently grasped at the resemblances between 1813 and OH 13 from Bed II at Olduvai to argue that neither of these fossils represents *habilis* (though he thinks 1470 probably does). We'll see later where this has led.

As the debate over the significance of ER 1470 and the other fossils progressed, it began to be drowned out by another controversy, this time over dating. The initial dating of the KBS tuff, using a new variant of the potassium-argon method, had given it an age of about 2.6 myr; but increasingly this date seemed to be contradicted by the evidence of the faunas from Koobi Fora. In particular, as early as 1971 a brief study by Basil Cooke of pig fossils from below the KBS tuff suggested to him that they matched pigs from Omo and Olduvai which were of much younger age than the KBS was supposed to be. This later turned out to be true for various other mammals as well. Since east Turkana and Omo, at least, were not far apart, at the same point in time their faunas should have been similar; that they apparently weren't made no sense, although ingenious attempts were made to rationalize this unusual situation. A rift began to emerge between Leakey's group at East Turkana and Clark Howell's contingent, still working at Omo. The Leakey side clung to the potassium-argon date; the Omo group to the evidence of the fossils, calibrated by their own series of chronometric dates. These suggested that the KBS tuff should be about 2 myr old. This growing disagreement also brought to light the fact that the geological framework developed in the early years of fieldwork at east Turkana had been severely deficient, as had been the documentation of exactly where each fossil had come from. Indeed, a definitive account of the geology there was not finally developed until well into the 1980s—ironically enough, by Francis (Frank) Brown, the geologist who had done so much to develop the geological framework for the Omo basin. The resulting inaccuracies in the dating of the hominid fossils naturally affected their interpretation, and, in particular, enhanced Richard Leakey's claim that very early *Homo* had existed in the region. The argument dragged on throughout the 1970s, eventually to be resolved by a combination of new chronometric and faunal studies in favor of a date of about 1.9 myr for the KBS tuff. This, of course, made ER 1470 almost exactly contemporaneous with *Homo habilis* from Olduvai Gorge, and to most observers this increased the likelihood that they belonged to the same species. Which in turn, of course, hastened the acceptance of *Homo habilis* as a real biological entity.

A major player in the resolution of the KBS tuff dating row had been Glynn Isaac, an archaeologist who had worked with the elder Leakeys and who joined the Koobi Fora Research Project in 1970 as archaeological director. When Isaac began work in east Africa in the early 1960s, Mary Leakey was the dominant influence in the archaeology of the earliest toolmakers; after all, the crude stone tools found in the lowest levels at Olduvai were at the time by far the world's oldest known. Mary had devoted much effort to classifying the various stone artifact types she recognized in the "tool kit" of the Oldowan toolmakers; but she was also interested in the nature of the sites at which such tools were found, and in what could be told from them about the activities of the early toolmakers. Some sites contained quite dense accumulations of animal bones as well as stone artifacts, and most were found in sediments that represented the margins of an ancient lake. Putting these facts together, Leakey concluded that such sites were favored spots to which early hominids had brought dead animals, which they butchered there with the stone tools. She thus regarded these places as "living sites," an interpretation that Isaac brought with him to East Turkana.

Turkana proved to have numerous artifact sites similar to those at Olduvai. In interpreting them, Isaac elaborated on Leakey's basic idea by introducing comparisons between the feeding behaviors of modern hunting and gathering humans and chimpanzees. Applying the ways in which humans differ from apes to sites where bones and artifacts were concentrated, he developed a model of early hominid behavior that saw these "campsites" as the focal points of a lifestyle involving the hunting and scavenging of animals (principally by males; females, encumbered by offspring, would mostly have gathered plant foods) and the transport of bits of their carcasses to a central place. Such transport was made possible by bipedal locomotion, which freed the hands for carrying. At campsites the carcasses were dismembered using stone tools made from materials brought in for the purpose, and the food was then shared among the members of the social group. This implied complex communication among group members, which might have amounted to language; and food sharing suggested some reciprocity in social relationships. Bipedalism, language, toolmaking, ranging around home bases, and complex sociality involving cooperation and the division of labor between the sexes, all thus fed back into a model which saw early hominids as incipient modern humans: a model that was warmly received by most of Isaac's colleagues since it fit well with the "man the hunter" view of the human condition that was popular among anthropologists at the time.

Later on, though, Isaac came to change his mind about nearly all of this. As a very careful scientist, he had always been concerned about the ways in which bone and tool concentrations might have come about; and there were certainly possible explanations other than human activity for the association of tools with animal bones. Further, he eventually began to realize that perhaps it was unwise to look upon early humans as too much like ourselves. These points were being forcefully made in the late 1970s by the iconoclastic American archaeologist Lewis Binford, and eventually Isaac set in motion an intensive reexamination of the archaeological record at east Turkana. In this the first goal was to determine whether the associations between bones and tools were indeed caused by hominid activities. If they were, what precisely were those activities? And if they weren't, what processes had caused the associations? By the time of his premature death in 1985, Isaac and his collaborators had made a good start on reanalyzing these questions. Most important, they had shown that the best of the sites at East Turkana did indeed bear the hallmarks of early human activity. For example, cut marks on the bones at some sites showed that animals had undoubtedly been butchered there using stone tools, while half or more of the stone flakes that had been used for such cutting exhibited the kind of wear that typically results from cutting meat.

From evidence of this kind, Isaac concluded that by 2 myr ago tool-using early hominids were cutting up the carcasses of a variety of large mammals. "One can only presume," he added dryly, "that they ate the meat that they cut." Further, certain body parts were disproportionately represented at various sites, and Isaac interpreted those parts as choice cuts that had been carried in after their owners had been killed or scavenged elsewhere. So some hint of the home base idea remained, although Isaac now admitted that this did not necessarily provide evidence of behaviors such as food sharing. Indeed, he suggested replacing the idea of "food-sharing" with the more noncommittal concept of "central-place-foraging." All in all, then, the pared-down picture of the earliest tool users that remained by the mid-1980s was considerably less dramatic than that which had prevailed a decade earlier—and a lot less human, too. But, as Isaac himself remarked, this process of simplification was necessary "to avoid . . . creating our origins in our own image."

As the archaeologists were refining their methods of analysis, the paleoanthropologists were continuing to turn up new fossils at Koobi Fora. During the 1974-1975 season yet another surprise came to light, in the form of a cranium dubbed ER 3733 that resembled nothing else yet known from

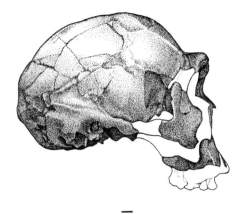
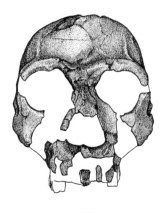

Side and front views of the cranium KNM-ER 3733, from the KBS Member, Koobi Fora, Kenya. Scales are 1 cm.   DM.

East Turkana. In his summary in *Nature* of the season's events, Richard Leakey merely remarked upon this specimen's evident similarity to *Homo erectus* from Zhoukoudian; but in an accompanying note written with Alan Walker, he was emphatic about its assignment to this species. Provisionally, Leakey and Walker reported a cranial volume of 800-900 ml, just reaching the lower limit of *H. erectus* (the eventual determination was 848 ml) and well above even the exceptional (though officially very much older) ER 1470. A date of between 1.3 and 1.6 myr was given for the cranium. Since such notable *Australopithecus boisei* specimens as ER 406 came from similar stratigraphic levels, Leakey and Walker took the opportunity to bash the proponents of the single species hypothesis with the absolutely undoubted, unequivocal coexistence of two distinct hominid species at East Turkana. It may seem remarkable that at this late date it was still necessary to belabor the fact that the single-species hypothesis was simply untenable; but it did indeed take this new evidence to put it to rest for good. Disappointingly, though, while Leakey and Walker pointed out that a new schema for human evolution was needed, they declined to provide one of their own.

Several more hominid fossils were found at East Turkana after this, including a slightly later but more robust braincase—ER 3883—that appeared to belong to the same species as ER 3733. But as the 1970s waned, the attention of the fossil collectors began to drift across the lake to the West Turkana region, and fieldwork in East Turkana became focused increasingly on the geology. As we've seen, deficiencies in the early geological studies had been thrown into glaring relief by the KBS tuff dating debacle, and new blood was brought in to rectify the situation. Notable among the new arrivals was the University of Utah's Frank Brown, who had worked with Clark Howell in the Omo until research there had to be suspended. In collaboration with his graduate student Craig Feibel, Brown eventually solved the problems of geological correlation between different areas of East Turkana using an ingenious new technique that allowed each of the datable tuffs in the area to be identified, wherever it was exposed, by its unique geochemical "fingerprint" (for every major eruption, even from the same volcano, emits material of slightly different composition). Using these datable tuffs as markers, Brown and Feibel were able to establish a uniform geological sequence for the entire Turkana basin, into which they were eventually able to tie the sites of discovery of most of the fossils that had been discovered. All the major sites fell in the time range of about 1.9 to 1.5 myr. Among the crania we've mentioned, the oldest were ER 1470 and 1813 at 1.9 myr; then came 407 and 1805, at about 1.85 myr; then 3733 at about 1.8 myr; then 406 and 732 at about 1.7 myr; and finally, 3883 at a little under 1.6 myr.

Younger than all of these was the mandible ER 992, at just over 1.5 myr. In 1975 this fossil had gained fame by being made the type specimen of the new species *Homo ergaster* ("work man") by the Australian systematist Colin Groves and his Czech colleague Vratja Mazak. Members of the Koobi Fora Research Project were appalled by this effrontery: how could mere outsiders presume to name *their* fossils? But this is the risk you take when you describe any distinctive fossil without naming it; and if the species is indeed a distinct one, the first name given to a member of it must take precedence. And in the end many paleoanthropologists—though not Leakey or Walker—did come to the belief that *Homo ergaster* was a good species. Something similar happened in 1976, when the Russian anthropologist V. P. Alexeev made ER 1470 the type specimen a new species that he called *Pithecanthropus* [*Homo*] *rudolfensis*. Of all of this, more later.

11

Hadar, Lucy, and Laetoli

While Richard Leakey's discoveries in the arid badlands of East Turkana were making his name a household word, a graduate student of Clark Howell's by the name of Don Johanson was preparing for equivalent fame in an equally inhospitable environment several hundred miles to the north. Johanson had accompanied Howell to Ethiopia in 1970 and 1971, and through members of the Omo expedition's French contingent he eventually met Maurice Taieb, a geologist whose field area lay in the Afar Triangle of northeastern Ethiopia. At its northern end the great East African Rift Valley splits in two, one branch heading northeast into the Gulf of Aden, and the other northwest along the Red Sea. The Afar Triangle marks the spot where the three rifting systems come together, and Taieb was studying the gological evolution of this unusual "triple junction."

During his surveys in the valley of the Awash River, Taieb had noticed rocks that he thought were probably Plio-Pleistocene, and which had abundant and well-preserved fossils eroding out of them. As a specialist in plate tectonics he wasn't interested in these fossils—but maybe Johanson was? Was he! Early in 1972, he joined Taieb, Yves Coppens, the pollen expert Raymonde Bonnefille and the American geologist Jon Kalb on a brief survey of the Afar, and at a place called Hadar they found a paleontologist's paradise: desert badlands oozing fossils that seemed by comparison with those from Omo to be around three million years old. Back in Addis Ababa, Ethiopia's capital, the group agreed to launch a full-scale joint expedition, which reached the field in the fall of 1973.

That first full field season at Hadar produced a remarkable find among a large haul of mammal fossils: the distal (far) end of a femur, plus the proximal (near) end of a tibia, which together made up the knee joint of a small hominoid. Short of a pelvis, perhaps, a more telling piece of the body skeleton couldn't have been found, for the knee tells you a great deal about locomotion. In a quadruped—an ape, say—the feet are held quite far apart, and each hind leg descends straight to the ground beneath the hip socket. In bipedal humans, on the other hand, the feet pass close to each other during walking so that the body's center of gravity can move ahead in a straight line. If this didn't happen, the center of gravity would have to swing with each stride in a wide arc around the supporting leg. This would be extremely clumsy and inefficient, wasting a lot of energy. So in bipeds both femora angle in from the hip joint to converge at the knee; the tibiae then descend straight to the ground. In the human knee joint this adaptation shows up in the angle—known as the "carrying angle"—that is formed between the long axes of the femur and tibia. The Hadar knee joint was clearly angled. At the end of the field season it was announced at a press conference in Addis Ababa that Hadar had yielded the knee joint of a bipedal hominid between three and four million years old, plus a piece of hominid temporal bone.

But this was merely a foretaste of what was to come. The next year, the field party at Hadar discovered, first, some hominid upper and lower jaws, and then "Lucy." Lucy, as all the world came to know in a remarkably short time, is the skeleton of a young adult female hominid, some 40 percent complete. She walked upright, as numerous details of her bony anatomy confirm, but stood only a litle over three feet tall. Her skull is extremely fragmentary, but it clearly had contained a brain in the ape size range (though given her diminutive stature it was probably a litle bigger in comparison to body size than an ape's). Her lower jaw is somewhat V-shaped, and while her molar teeth are quite humanlike, the front premolars are not bicuspid like ours. But what was most breathtaking about Lucy was the combination of her age and her completeness. Up to 1974, the earliest reasonably complete hominid skeletons known were those of Neanderthals, close relatives of *Homo sapiens* and under 100,000 years old. As documentation of earlier stages in human evolution, only isolated bones were available. The only pre-Neanderthal hominid specimen that came even remotely close to Lucy in completeness was Broom's *Australopithecus africanus* pelvis from Sterkfontein, with its associated partial femur and some vertebrae. And while these remains were sufficient evidence on which to conclude that their owner had been bipedal, that was about all that could be said. Lucy, on the other hand, was complete enough to provide a pretty comprehensive picture of the kind of individual she had been. And she was also a good half-million years or more older than the fossil from Sterkfontein.

A year after the initial anouncement of a 3 myr old hominid from Hadar,

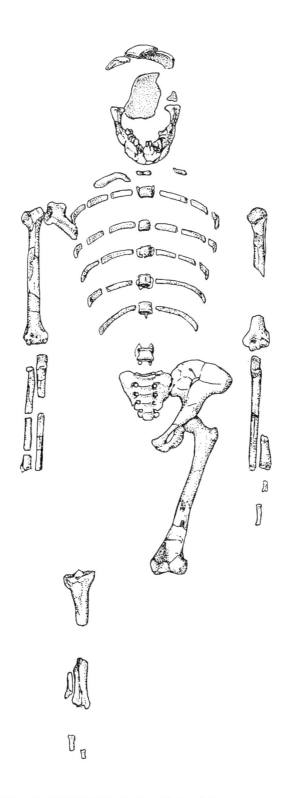

The skeleton of "Lucy" (NME-AL 288-1), from Hadar, Ethiopia.   DS.

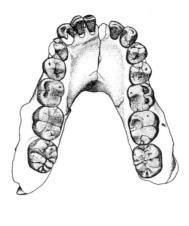

Lower jaw with almost complete set of teeth (NME-AL 400-1a) from Hadar, Ethiopia. Scale is 1 cm. DM.

then, Lucy provided clinching evidence to show that human precursors had indeed been up and walking on their hind legs at that remote point in time. But what was Lucy? What species did she and the other Hadar fossils represent? Just before finding Lucy, Johanson had been visited in the field by Richard Leakey and other members of the Koobi Fora group. With modestly proportioned bony parts and a rather *Homo*-like balance in size between the incisor and molar teeth, the jaws that had already been found clearly did not belong to *Australopithecus boisei*, with its massive jawbones, tiny incisors and massive chewing teeth. So Leakey, fresh from his discovery of ER 1470, quickly and predictably concluded that they must belong to an early species of *Homo*. By his own admission Johanson himself was already toying with this conclusion. On the other hand, it was clearly meaningless to consider the diminutive and tiny-brained Lucy as a member of our own genus. Thus, in their preliminary description in *Nature* of the 1973 and 1974 Hadar hominids, Johanson and Taieb concluded that two or three species were represented among them: a very primitive form of *Homo* in the shape of the isolated upper and lower jaws; and, in the shape of Lucy and the 1973 knee joint, something else. Just what remained to be determined, although Johanson and Taieb felt that these remains bore comparison with *Australopithecus africanus* from South Africa. And finally, they noted resemblances to robust *Australopithecus* in a temporal bone (part of the skull wall) found at the end of the 1973 field season.

The undoubted if as yet obscure importance of the Hadar hominids made it imperative to get them securely dated. To help with this task Johanson, who had just been appointed to a curatorship at the Cleveland Museum of Natural History, recruited the services of James Aronson, a potassium-argon dating specialist at Case Western Reserve University in the same city. Taieb had already located a datable lava flow at Hadar, plus some thin tuffs, and in 1974 Aronson visited Hadar to collect datable samples. Aronson was able more precisely to fix the position of the lava flow in the Hadar stratigraphy, and following his return to Cleveland dated it at over 3 myr, confirming preliminary dates on a sample collected earlier. Nonetheless, because of suspected weathering of the lava samples (which would have resulted in the loss of accumulated argon and thus an underestimate of the true age), it remained possible that the lava was in fact somewhat older. At the top of the section, a capping date of 2.6 myr was obtained on a tuff; this resulted in an estimated age for Lucy of about 2.9 myr, while the isolated jaws and the knee joint were older. For a variety of reasons uncertainties about the precise dating at Hadar lingered until the early 1990s, when it was found that all the many hominid fossils known by then came from a rather short span of time, between about 3.2 and 3.4 myr. Lucy was the youngest among them. This made Lucy older, and the older specimens somewhat younger, than had generally been thought.

The 1975 field season saw Ethiopia in political turmoil. Haile Selassie had been overthrown in 1974 and his powers assumed by a Marxist military dictatorship, whose structure took some time to emerge. Nonetheless, Johanson and his colleagues managed to get back to the Afar, and they hit the jackpot yet again. This time it was the "First Family," an unbelievable trove of some two hundred early human fossils, all jumbled up next to one another in the sediments. As in the case of Lucy, nothing resembling this had ever been found before. Site 333, as the locality at which these fossils were found became designated, eventually yielded the fragmentary remains of thirteen individuals, male and female, adult and juvenile. The means by which the remains of all these hominids came to be buried together in the sediments has never been definitively figured out, but one suggestion was that all may have been caught together in a flash flood. If this is the case, then it is likely that all belonged to the same social group. And although one group doesn't make much of a statistical sample, if most of its members were represented among the fossils, this would tell you something about social group size—and thirteen individuals of all ages is not out of line with expectation. More important, though, if all belonged to the same social group, then all belonged to the same species. Paleontologists almost invariably have to rely on inference in determining whether different fossil individuals in fact belonged to the same species; but here, potentially, is a demonstration of this by a totally different means.

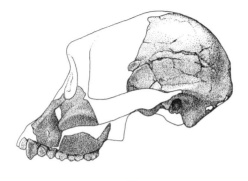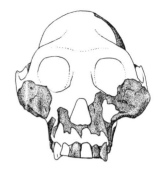

Side and front views of reconstruction of the cranium of Australopithecus afarensis, *made on the basis of various unassociated fragments from Hadar, Ethiopia. Scales are 1 cm.*

DM.

Disappointingly, however, there is no unanimity on this matter of burial; some claim that the fossils acumulated sequentially over some period of time, and the debate seems set to continue. This is particularly a shame because whether or not the Hadar hominids represent a single species continues to be an especially contentious subject in paleoanthropology.

Despite continuing instability in Ethiopia, the Hadar team returned for further fieldwork in late 1976. Aside from more hominid fossils, notably many new ones from the First Family site, these researches resulted in the discovery by the archaeologist Hélène Roche of simple basalt tools dated at about 2.5 myr. Who had made these remarkably ancient tools remained problematic: no hominid fossils were known from this point in time, and in any case associating early stone tools with their makers has been a perennially tricky issue in paleoanthropology. Sadly, though, it proved impossible to follow up on this find. For, as the result of another military coup in Addis Ababa, the end of the 1976-1977 field season marked the effective cessation of work at Hadar: work which wouldn't resume on any major scale for well over a decade.

It's appropriate to note here that the work at Hadar was not the only paleontology being pursued in Ethiopia during the mid-1970s. In 1976 a group led by Johanson's former collaborator Jon Kalb found the face of a massively built adult hominid in the Middle Awash region of the Afar Depression. This specimen was found lying on sediments belonging to the Upper Bodo Beds. These contain abundant vertebrate fossils and Acheulean tools, and they were dated by Kalb and his associates to the middle Pleistocene, perhaps about 400 kyr ago (with a very wide margin of error).

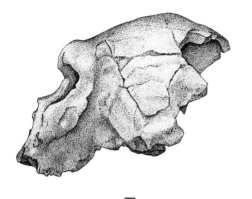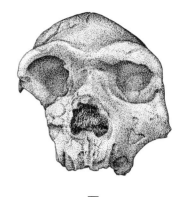

Side and front views of the partial cranium from Bodo, Ethiopia. Scale is 1 cm.   DM.

However, it's not possible to say for sure that they are the actual sediments from which the fossil was derived—which is particularly unfortunate because the fossil belongs to a distinctive group that is itself exceptionally poorly dated. In Africa, the closest comparison is with the famous "Rhodesian Man" cranium from Kabwe, Zambia. Judiciously enough, the Kalb team found the Bodo face to be less archaic than Asian *Homo erectus* or Olduvai Hominid 9, but to be more so than the Omo Kibish crania that Richard Leakey's group had discovered a few years earlier. We'll come back to what that means later; meanwhile, suffice it to note that during a brief survey of the same area in 1981, a group led by the veteran African archaeologist J. Desmond Clark recovered a couple of hominid fragments from very much older sediments dated at about 4 myr ago. These fossils consisted of a partial frontal bone and part of the proximal end of a femur belonging to a habitual upright biped. The frontal fragment was said by Clark and his colleagues to resemble an immature specimen known from the Tanzanian site of Laetoli, whither the focus of this story now shifts.

Laetoli, we've seen, is a place about twenty-five miles southwest of Olduvai Gorge that was first visited by Louis and Mary Leakey in 1935. At that time they recovered an isolated lower canine tooth that Leakey thought belonged to a monkey, but which eventually turned out to have come from the jaw of an early hominid. There are several square kilometers at Laetoli over which sediments of Pliocene age are exposed, but there is much more vegetation around than there is in the gorge at Olduvai. This makes it much less satisfactory from the fossil collector's point of view, so the elder Leakey rapidly turned his attentions to Olduvai, where he remained even after the

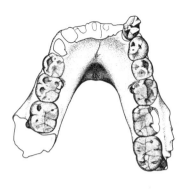

The LH 4 mandible from the Laetolil Beds, Laetoli, Tanzania. Type specimen of Australopithecus afarensis. *Scale is 1 cm.*   DM.

German explorer Ludwig Kohl-Larsen turned up a human upper jaw fragment at Laetoli (or Garusi as he called it, for the river valley in which it lies) some four years later. In 1950, as we've seen, Kohl-Larsen's compatriot Hans Weinert made this fragment the type specimen of the new species *Meganthropus africanus*, considering it allied to a problematic jaw fragment from Java that had been named by Franz Weidenreich in 1945. But it was not until five decades had elapsed after her first visit that Mary Leakey was to return to Laetoli to undertake prospections of any magnitude. These field researches lasted from 1974 to 1981 and yielded some thirty early hominid fossils, ranging from isolated teeth and jaw fragments to two quite well-preserved adult lower jaws, one adult (Laetoli Hominid 4) and one juvenile (LH 2). Bits of a juvenile skeleton were also recovered. Another find was of a much later and reasonably complete human skull (LH 18, known as the Ngaloba skull after the deposits in which it was found); this is somewhat archaic in appearance and has an estimated brain volume of 1,200 ml. Potassium-argon dating on tuffs revealed that the two jaws were between 3.6 and 3.8 myr old; the Ngaloba specimen is probably in the region of 150 kyr old.

The real jewel of Laetoli, however, was a number of footprint trails initially discovered by the paleoanthropologist Andrew Hill as he dove to the ground to avoid a lump of elephant dung flung at him by a colleague. These trails were preserved in very fine volcanic ash that was puffed out some 3.7 million years ago by the nearby volcano Sadiman. After this pow-

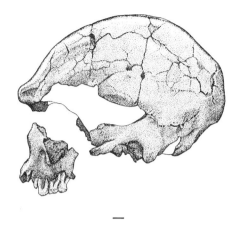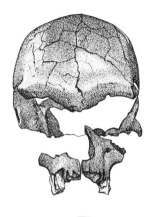

—　　　　　　　　　　　　　　—

Side and front views of the LH 18 partial cranium from the Ngaloba Beds, Laetoli, Tanzania. Scales are 1 cm.　　　　　　　　　　　　　　　　　　　　　　　　DM.

der had settled in a thin layer on the landscape a light rain fell, turning it to something like wet cement. Before it dried out and hardened various birds and mammals wandered over it, leaving their prints behind. Among the mammals were some early hominids, and, incredibly, at one place (Site G, excavated in 1978 and 1979) the preserved footprints of a pair of such creatures who had walked side by side were preserved over a total distance of about eighty feet. After the prints had been made the volcano erupted once more, emitting more ash, which covered and protected them until they were exposed by erosion for the Leakey group to find. This was, of course, a quite remarkable and unprecedented discovery. Most fossils provide direct evidence of bony or dental anatomy only. Any inferences about how their possessors had behaved—walked around, for instance—are just that, inferences. But here at Laetoli early human behavior itself was fossilized! Of course, these were not the only footprints of ancient humans ever to be found: others are known, for instance, from several Ice Age caves in Europe. But all the other fossil human footprints known are those of members of our own species, *Homo sapiens*, dating back at most a few tens of thousands of years. The Laetoli prints, in contrast, reflect the origins of the human lineage; they are the only prints so ancient that the creatures who made them might conceivably not have been fully bipedal—who might not have walked just as we do.

Of course, the world being the imperfect place it is, full agreement over the implications of the the prints was long in coming—has not, indeed, been achieved yet, although nobody disputes that they are the tracks left behind by upright bipeds. Similar disagreement surrounds the interpretation of the hominid fossils from Laetoli, although initially they assumed a low profile. The jaws and teeth found at Laetoli were given for description to Tim White, a student of Milford Wolpoff's who had worked briefly at East Turkana before falling out with Richard Leakey, and who later spent the 1978 field season (when the majority of the prints were excavated) at Laetoli before falling out with Mary Leakey. In 1977, White produced a meticulous description in fine print of the Laetoli hominids discovered in 1974 and 1975, without making any comment whatever about their affinities. Descriptions of the later finds followed in 1980. I recall colleagues complaining at the time how unreadably boring these descriptions were, but—as he would shortly prove—this was hardly due to any shortage of imagination on White's part; he was simply hewing closely to a policy that the Leakeys were by then imposing on all their collaborators.

So what was this early biped of Laetoli, effectively the earliest hominid known? The first attempt to answer this had to await a collaboration between White and Don Johanson. Between them, these two researchers were responsible for the description of virtually all the hominid fossils known then from the period between about 4 and 3 myr ago, and it was in a way natural that the two should have ended up collaborating on an analysis of these fossils, despite the fact that the two researchers were of very different temperaments. In any event, in the summer of 1977 Johanson asked White to bring casts of the Laetoli fossils to Cleveland for comparison with those from Hadar. The question was obvious: Did the hominids from the two sites belong to the same species?

Disagreement between Johanson and White on this point might have been expected; after all, Johanson had already claimed in print that several species might be represented among the Hadar fossils, while White had trained under Milford Wolpoff, the guru of the single species philosophy that rejected the idea that more than one hominid species could exist at any one point in time. And by Johanson's account disagreement there was, at least to start with. One problem was that there was an enormous range in size between the biggest and the smallest individuals represented among the Hadar fossils. And there were certain differences in morphology, too; for example, while the tiny Lucy had very diminutive front teeth (hence the rather V-shaped jaw), bigger jaws from Hadar had relatively larger front teeth. The size differences might simply have been the result of very substantial sexual dimorphism within a single species, the larger fossils representing males and the smaller ones females. In that case, the shape differences would very likely have been no more than passive consequences of size discrepancies. The influence of size on shape had, after all, been documented long since among living organisms of all kinds. Alternatively, both size and shape differences might have been accounted for by the presence of more than one species in the sample. White favored the single-species view, despite the huge disparity in size between the largest individuals and the smallest ones. To begin with, at least, Johanson preferred the two-species interpretation he had already espoused, with the larger specimens representing some primitive species of *Homo*, and Lucy and her like something else.

The first conclusion the pair agreed upon was that all the Hadar hominids were distinct both from apes and other known hominids. Then it became evident to them that a virtually complete gradation in size from large to small was present within the sample, so that the question of shape differences became paramount. In the end they satisfied themselves that the shape differences were adequately accounted for by allometry plus the differences that always occur between different individuals of the same species. At Hadar and Laetoli, between about 3.7 and 3 myr ago, they thus saw evidence of a single species of hominid that was unlike any other. This species was fully bipedal, and males were much bigger than females (somewhat as among gorillas today), though even males probably didn't stand a lot taller than four and a half feet. To judge from the pronounced muscle attachment scars on the bones, both sexes had been powerfully built. Their arms were longer in proportion to the legs than ours are today, and their hands were like human ones in most details, though they were a bit longer and more curved. Their torsos tapered upward. Their brains were small, in the chimpanzee range, and, again like those of apes, their faces were large and projecting. Their toothrows converged a little toward the front, forming an arcade that was neither parabolic, as in modern humans, nor parallel-sided, as among living apes. The incisors were large, though not extraordinarily so; and while the canines were conical, as in apes, they resembled those of humans in their reduced size. The molar teeth were large and relatively flat. These distinctive hominids predated the earliest stone tools known from Hadar, and almost certainly they were not toolmakers.

Here, then, Johanson and White had evidence for a new species of hominid, one earlier than any other known. But a species of what genus? There were only two choices on offer: *Australopithecus* and *Homo*, and nobody who didn't wish to be hooted out of the profession would have dreamed of creating a new genus. Based more than anything else on the age of their new creature, they decided that what they had here was a stem hominid, ancestral to all subsequent human species. And if this ancestor had given rise to *Australopithecus* as well as *Homo* species, they reasoned, it couldn't itself be *Homo*. The logic of this particular proposition is not entirely clear (for if a genus located at the fork of a V can be given the name of one of the two at the ends of the branches, it can equally well be given that of the other); but there's no doubt that it was the more palatable choice. For to call this creature a member of *Homo* would effectively be to make our own genus equivalent to the entire family Hominidae.

So the genus *Australopithecus* was selected. What was the new species to be named? Given that the fossils came from two widely separated sites, Johanson and White wanted to bind them together symbolically. This they achieved by making the Laetoli mandible LH 4 the type specimen of their new species, and by calling that species *Australopithecus afarensis*, for the Afar region of Ethiopia that had yielded most of the specimens assigned to it. This, though legal, was not in fact very good procedure, for it remains at least potentially possible that someone will come along and demonstrate that the Laetoli and Hadar materials in fact belong to different species—as at least one scientist, indeed, already believes she has. Should this be conclusively proven, the name *afarensis* will have to stay with the Tanzanian type specimen. The chosen name did, however, reflect the strength of Johanson's and White's conviction that only a single species existed at the two sites; and most criticism of this idea, when it came, focused not so much on differences between the two sites as on whether more than one species might be recognized among the multitude of fossils known from Hadar.

The single-species idea conflicted, of course, with the views that Richard Leakey was propounding on the basis of his Koobi Fora fossils—especially given the very early date that he was still assigning to the KBS tuff. Although Leakey had never been very specific about his views on the pattern of early human evolution, his publications showed clearly that he believed in the separation of the *Homo* and *Australopithecus* lineages from the very earliest times; to see them converging as claimed by Johanson and White was unpalatable to say the least. The first rumblings of trouble came in 1978, when Mary Leakey attended the meeting in Stockholm at which Johanson made the initial public announcement of *Australopithecus afarensis*. Mary had been intending to discuss the Laetoli hominids, and according to Johanson she was affronted to hear him talking about "her" specimens, even though they had been decribed in print and thus were in the public domain. Further, the implications of what he was saying about them were hardly congenial to her. By Johanson's account she had earlier agreed to join him, White, and Yves Coppens as coauthor on the publication describing the new species (with the proviso that it not suggest that any *Australopithecus* was ancestral to *Homo*; this was achieved by eliminating any discussion of the affinities of the new species). Now, however, she demanded that her name be taken off the article, which at that stage could be achieved only by destroying the entire print run and publishing a new version. Personal relationships between the Leakeys and the Johanson-White axis became decidedly cooler.

While all this was going on, Johanson and White prepared an interpretive paper on *A. afarensis* that laid out their ideas on the place of this species in human evolution. Published at the beginning of 1979 in the influential journal *Science*, the article made headlines in the popular media as well as waves in the profession. In it the pair discussed various alternative possible arrangements among the hominid species that they recognized, and they plumped for a simple forking scheme whereby the stem species *A. afarensis* gave rise, some time after three myr ago, to two lineages. One of these led from *Homo habilis* through *H. erectus* to *H. sapiens*; the other led from *A. africanus* to *A. robustus*, which became extinct at about 1 myr ago. They did not recognize the Leakeys' species *A. boisei* as distinct from South African *A. robustus*, and they claimed that tendencies toward the robust condition were already detectable in *A. africanus* compared to the more primitive *A. afarensis*.

This (or, to be quite frank, virtually any other) interpretation was bound to cause an outcry, and not simply from the Leakeys. One might well, indeed, have expected the response to have been even more shrill than it actually was, but it was nonetheless hardly muted. Some critics were adamant that what Johanson and White had were actually East African representatives of the species *A. africanus*, already abundantly represented from South Africa though only questionably known in eastern Africa by a few isolated teeth from Omo. Ironically - because his own new species, *Homo habilis*, had suffered the identical criticism—South Africa's Phillip Tobias was foremost among the critics. Indeed, Tobias announced during the Stockholm conference at which *A. afarensis* made its debut that the Laetoli and Sterkfontein fossils merely represented subspecies (geographical variants) of *A. africanus*. He repeated this assertion in a long paper published in 1980, in which he claimed that it was the expanded species *A. africanus* that was the progenitor of all later hominids, including the robust australopithecines on one hand and *Homo* on the other.

At about the same time Todd Olson of New York's City College reviewed the same question and concluded that the bulk of the Hadar specimens and those from Laetoli in fact belonged with the robust australopithecines—which he allocated to Broom's genus *Paranthropus*. Most specifically, he found *Paranthropus* characteristics in skull base fragments from Hadar. Accepting the priority of Kohl-Larsen's species name for the Laetoli jaws and teeth (and by extension, for the larger Hadar specimens), he allocated these East African fossils to the species *Paranthropus africanus*. On the other hand, he claimed, the smallest of the Hadar specimens, notably Lucy, were different. These he placed in a primitive species of the genus *Homo* and, like John Robinson, he extended this genus to embrace the gracile South African material. Thus in Olson's view the small Ethiopian fossils, plus those traditionally placed in *Australopithecus africanus*, all belonged to species of *Homo*.

Another paleoanthropologist who now saw at least two species among the Hadar fossils was Yves Coppens, whose students Brigitte Senut and Christine Tardieu had been studying the postcranial remains. Like Olson, these researchers arrived at the conclusion that two (or maybe more) species of hominid were represented at Hadar. However, Coppens saw Lucy, along with the bulk of the material, as a member of *Australopithecus afarensis*. It was various other arm and leg bones that they saw as belonging to a primitive species of *Homo*. And Richard Leakey, of course, continued to find evidence in the sample for both *Australopithecus* and *Homo*, although he offered no detailed reasons for this.

The argument continues; but the Johanson-White interpretation that only a single hominid species occurred at Laetoli and Hadar quite rapidly carried the day among most paleoanthropologists, at least as a working hypothesis. Indeed, *Australopithecus afarensis* entered the accepted pantheon of ancient human precursors a good bit more rapidly than had any fossil human species previously named. But fossils are not, of course, merely static objects that merely sit gathering dust in museum drawers when we are not trying to find out from them to which other fossils they are most closely related. In the final analysis they are the only witnesses we have to a long and dynamic and eventful story: a story of creatures struggling to survive and to perpetuate themselves within an environment that tends—wherever on Earth—to be both dangerous and unpredictably changeable. And how our ancestors coped with such vagaries constitutes a significant part of the evolutionary story of our lineage.

At the time when *A. afarensis* was described, much common wisdom ascribed the adoption of bipedalism to the need to free the hands to make tools and to carry them and other objects around. But the discovery that this upright biped was already up on its hind feet a good million years before stone tools appeared in the archaeological record obviously meant that some rethinking was due. The most elaborate of this rethinking was done by Owen Lovejoy, of Kent State University, to whom Johanson had entrusted the description of much of the hominid postcranial material from Hadar. Lovejoy's analysis of the Lucy skeleton and other fossils convinced him that here was not only an upright biped, but a creature that was very efficiently adapted to an upright striding gait. He found, for example, that Lucy's restored pelvis not only showed all the hallmarks of our own, but additionally had more widely flaring ilia (the "blades" of the pelvis). In combination with a longer neck to the femur, this attribute improved the mechanical advantage of the muscles that stabilize the hip in the upright position. And this was possible, he reasoned, because the small-brained *A. afarensis* simply didn't need to make the compromises that are necessary in modern humans to allow the passage through the birth canal of a large-brained newborn. Lovejoy thus concluded that in *A. afarensis* upright bipedalism was in fact more efficient than it is in us.

This conclusion hasn't gone unchallenged. Their studies of the bones of the upper and lower limbs suggested to Senut and Tardieu that joint mobility was greater in *A. afarensis* than in modern humans, implying greater climbing capabilities. Bill Jungers of the State University of New York at Stony Brook pointed out that while Lucy's arms were proportionately not much longer than those of modern people, her legs were shorter, which would favor climbing. Russ Tuttle of the University of Chicago found that the length and curvature of the bones of the hands and feet suggested a strong grasping, hence climbing, capability. Henry McHenry of the University of California at Davis noted high mobility in the wrist joint, which carries similar implications. Putting all this together, Jungers and his Stony Brook colleagues Randy Susman and Jack Stern concluded that whereas *A. afarensis* was undoubtedly bipedal while on the ground, it probably spent a good deal of time in trees. They felt it likely that at night these early hominids sheltered from predators in trees, and probably also foraged there in the daytime, too.

How you look at all of this depends, of course, on which characteristics you think are most important in determining habitual behaviors. Nobody disputes that *A. afarensis* was descended at some remove from a largely tree-dwelling ancestor, although we have to bear in mind that modern great apes all spend more or less time on the ground. Obviously, terrestrial bipedalism was not acquired overnight in its full anatomical splendor, so we would expect newly bipedal hominids to show some evidence of their arboreal ancestry in their skeletal structure. And if we thus expect to find a mosaic of terrestrial and arboreal characteristics in the first habitual bipeds, which of those characteristics should we regard as most informative about behavior? It's close to certain that the newly acquired (in this case, terrestrial) features of *A. afarensis* reflect behavior—for why else would they have become established? The big question thus reposes on how much the ancestral tree-climbing capacity was actually used.

The environment of *Australopithecus afarensis* at Hadar consisted of a fluctuating mosaic of riverine gallery forests and more open savanna habitats, and it presumably moved through both (the arid grassland environment in which the Laetoli hominids left their tracks was almost certainly not typical of where they found the bulk of their sustenance). What's more, though robust, *A. afarensis* was small-bodied and, being bipedal, it wasn't very fast. Presumably, then, this hominid was pretty vulnerable to open-country predators, and as a reasonably accomplished climber it would hardly have refrained from using trees for shelter, particularly at night. Further, while tree-borne fruits would have been within its grasp, as far as we know it did not use tools. And this would have limited its access to many of the resources—roots, bulbs, tubers and so forth—that were potentially available on the savanna. In the deadly serious game of survival it's highly unlikely that *A. afarensis* would not have used every resource at its disposal, so on balance a behavior pattern that combined its climbing abilities with its newfound bipedal capacity seems probable. And since the anatomical structure typified by Lucy seems to have endured for at least a couple of million years, this was clearly a successful behavioral strategy.

The exact interpretation of the functional anatomy of Lucy and her kin is still disputed. It's claimed, for instance, that *A. afarensis* could not have fully extended its knee as we do, while the head (the weight-bearing portion) of the femur is much smaller than it is in us, suggesting a less complete adaptation to upright posture. Nonetheless, the essentials of bipedal locomotion were undoubtedly there, and few would doubt now that bipedalism was the primordial hominid adaptation. Which leads to the obvious question: Why? Owen Lovejoy thought he had the answer. In 1981 he published a paper arguing that, since the anatomically and behaviorally complex transition from ape-style quadrupedalism to upright posture could not have taken place in a single fell swoop, there must have been some countervailing advantage that increased the reproductive success of the neither-fish-nor-fowl early hominids. By themselves the females couldn't do much to increase their rate of reproduction since they were already hampered by offspring that took years to become independent. But they could achieve that result by co-opting males into the feeding of the family. Unencumbered males were better able to roam around the landscape, and if bipedal would have had their hands free to carry food home. However, they could benefit reproductively from such an arrangement only if the offspring whose survival prospects they thereby increased were actually their own, which conveniently meshed with the females' obvious interest in having a permanent mate to depend upon. In this way Lovejoy ingeniously wove bipedalism, food-carrying, and ranging around home bases into a scenario that also involved the development of pair bonding and fidelity among early hominids. In turn, through the system of permanent sexual signals that reinforced it, this bonding accounted for the marked but famously hard to explain secondary sexual differences (prominent breasts and facial hair, for instance) that occur between human males and females.

Predictably, Lovejoy's ideas ran into a lot of criticism on a variety of grounds. But they had the virtue of opening up the question of the origins of human bipedalism to an intensive reexamination. There was much discussion, for instance, of the energetics of bipedal locomotion, both in the style of humans and in that of modern great apes, all of which have a propensity for carrying their trunks upright at least under certain circumstances. Peter Rodman and Henry McHenry of the University of California at Davis, for example, showed quite elegantly that while human bipedality is indeed inefficient compared to the quadupedalism of a committed terrestrial mammal such as a horse, it is actually relatively efficient compared to terrestrial quadrupedalism in the ape style (which necessarily represents a compromise with locomotion in the trees). If an ape found itself having to cover long distances on the ground (as might happen as the forests in which it lived were fragmented by encroaching grassland), bipedalism might indeed be the most efficient form of locomotion for it to adopt. According to Rodman and McHenry, there was no need to invoke a fancy behavioral advantage to explain the transition from hominoid-style quadrupedalism to bipedalism, simply because it could have made good energetic sense all by itself: there was no unbridgeable energetic gulf to be crossed between hominoid quadrupedalism and hominid bipedalism.

This kind of argument fitted well into a burgeoning scrutiny of the role of environmental change in the adoption of hominid bipedalism—especially after it was found that the origin of the human family probably coincided reasonably closely with a drying episode in Africa. During this event forest cover on the continent shrank considerably and was replaced by grasslands over wide areas. The most interesting recent speculations have centered on the changed physiological demands that faced the protohominids as they began to emerge into this new environment while their ape cousins remained confined to the steadily diminishing forests. For example, the English physiologist Pete Wheeler recently explained how problems of body temperature regulation, together with a shortage of drinking water, must have challenged these human precursors—and how bipedalism must have helped to meet that challenge.

Perhaps the most critical physiological problem facing any savanna-living mammal is cooling the brain, an organ that is highly sensitive to any overheating. Most savanna mammals have special mechanisms devoted to this function, but, as forest dwellers, most primates don't. The only means available to the first hominids for cooling the brain was thus to keep the whole body cool—and one way of doing this was to minimize the incoming heat load imposed by the tropical sun. This is precisely what an upright posture achieves, by minimizing the area of the body exposed to the sun's direct rays. What's more, bipedalism raises the body far off the ground, where it can be cooled by the wind. In this way heat is lost by convection as well as by the evaporation of sweat—especially if the skin is not insulated by the dense hairy coat that bipedal posture makes it advantageous to shed. It is also virtually certain that early savanna-dwelling hominids would have had to range fairly widely to find food, and Wheeler has calculated that at slow speeds human bipedalism demands less energy than ape quadrupedalism. This means that less internal body heat is generated as a by-product of energy production. With less heat generated internally and less absorbed from the environment, and a larger proportion of the body's surface area sheltered from the sun's direct rays and thus available for radiative cooling, body temperature regulation in the tropic environment ceases to be a critical problem. Wheeler notes that the arboreal ancestors of the first human bipeds were almost certainly not committed quadrupeds; rather, they were semiarboreal generalists which already had a propensity to hold the trunk upright. When the forests in which they had lived began to fragment and to be replaced by sun-scorched grasslands they had a number of options open to them as they began to exploit the new environment; the physiological advantages of upright posture may have been enough to tip the balance in favor of bipedal locomotion.

Later humans developed specialized means of cooling the brain, among them a "radiator" composed of tiny veins in the scalp and face. Dean Falk of the State University of New York at Albany points out that this mechanism (whose efficacy is debated) is lacking in living apes. What's more, to judge from features of the inside of the braincase that are associated with such cooling, it was also absent in the hominids from Hadar and in the robust australopithecines. Falk claims, though, that the pattern of cranial blood circulation differs in the one (juvenile) specimen from Laetoli that bears on this issue. In her view, this may place the Laetoli and Hadar fossils not simply in different species but in different lineages, the former lying on the way to gracile *Australopithecus* and *Homo*, and the latter giving rise to the robust australopithecines. This is not an interpretation that has attracted a lot of support, but it is the kind of finding that raises questions about how we interpret the fossil data at our disposal. And since the heyday of fossil hominid discoveries in Kenya and Ethiopia began in the earlyand middle 1970s, there have been twin revolutions in the ways in which paleoanthropologists—and paleontologists in general—view both the evolutionary process and the fossil record of evolutionary history. We'll look at those revolutions in the next chapter.

12

Theory Intrudes

Quite apart from the extraordinary additions to the hominid fossil record which were made during the 1970s, that decade was a period of great excitement and ferment in evolutionary biology. For years paleontologists had labored mightily to fit the evidence provided by their fossils into the framework of stately change dicatated by the New Evolutionary Synthesis; and by around 1970 some of them were coming to find that fit increasingly uncomfortable. The Synthesis, as you'll recall, elegantly explained all evolutionary phenomena in terms of the gradual accretion of genetic changes in evolving lineages, all under the guiding hand of natural selection. In turn this implied that species, while discrete units in space, should lose definition in the dimension of time. Species were, in fact, viewed as nothing more than arbitrarily defined segments of evolving lineages which, if they didn't die out leaving no descendents, would inevitably evolve into something else. Time and anatomical change were thus thought to be more or less synonymous. The implication of this was that the fossil record should consistently show smooth intergradations from one species to the next; inconveniently, however, it too often didn't. Species, it has turned out, tend to appear rather suddenly in the fossil record, to linger for varying but often very extended periods of time, and to disappear as suddenly as they arrived, to be replaced by other species which might or might not be closely related to them. For a long time—indeed, since Darwin himself—this failure of the fossils to accord with expectation was explained away by the famous incompleteness of the record. But as the years passed and more and more fossils were found, the predictions of the Synthesis became increasingly out of sync with what was actually there. The time was evidently ripe for a reappraisal of paleontologists' expectations from theory—and thus of the theory itself.

The doctrines of the Synthesis had not, of course, been accepted by everyone. But a concerted attack on its assumptions came only with the 1970s. The first signs of effective opposition came in 1971, from my American Museum of Natural History colleague Niles Eldredge, who had been studying the evolution of a group of trilobites (ancient sea bottom-dwelling invertebrates) that is abundantly represented in the rocks of upper New York State and the Midwest. Eldredge had noticed that among the trilobites of interest to him there was a marked lack of evolutionary change; indeed, in the Midwest during an 8 myr span (since reduced to 6 myr) there was only one anatomical change that might be interpreted as heralding a new species. This was a decrease in the number of rows of lenses in the insect-like compound eye, from eighteen to seventeen: earlier sites had the eighteen-row form, later ones the seventeen-row kind. A similar pattern obtained in New York; but at one site there, which preserved a single instant in that time, fossil trilobites of both kinds were represented. And this site was very much earlier than the transition from the one to the other in the Midwest. This suggested to Eldredge that a short-term speciation event (the emergence of a new species) had taken place in what is now New York, and that in the Midwest things had stayed as they were for millions of years until an environmental change had permitted the new species to invade and replace its forerunner there. The message was clear: the trilobite record in this part of the United States was overwhelmingly one of stasis (stay-as-you-are) rather than of continuous change. This interpretation, which took the fossil record at its face value rather than as an inadequate reflection of the past, was of course totally at odds with the expectation of gradual change. And rather than try to rationalize what he saw as an example of the legendarily incomplete record, Eldredge stuck with his idea of stasis and invoked allopatric speciation to explain the pattern he discerned.

Allopatric speciation is an old concept that had been most notably elaborated by Ernst Mayr to explain how one species could give rise to another. Noting that what fundamentally distinguishes members of closely related living species is their inability to produce viable and fertile offspring (an inability which, of course, paleontologists can't observe in the fossil record), Mayr had proposed that new species arise when a geographic barrier of some kind (a river, perhaps, or a mountain range, or a desert) disrupts the territory of a widespread species. The newly separated populations, formerly part of one big interbreeding whole, will then be unable to exchange genes. Since mutations and other genetic differences will continue to arise in both populations, the two will start to diverge, a process that will ultimately prevent effective reproduction between their members even if they come into contact again. In this way, two species emerge where there was only one before. Because there's less genetic inertia in the gene pools of small populations, Mayr felt it most likely that speciation and change would occur in small peripheral isolates, rather than in big chunks of the parental population. And if, by the way, you're asking just how Mayr squared this model with the grand ideas of the Synthesis that he did so much to establish, the answer is that he never quite did.

We'll look again at species and how they form and can be recognized by zoologists and paleontologists; suffice it for the moment to note that in geological terms allopatric speciation takes but an instant. It's also independent of long-term anatomical change. Eldredge interpreted his mixed sample of trilobites from that New York locality as a population caught in the process of allopatric speciation (for clearly this wasn't something that was taking place throughout the range of these trilobites). But Darwin and his followers had contended that the origin of species lay in adaptive modification, a slow process that took place over vast spans of time. And although the framers of the Synthesis had been careful to acknowledge that rates of evolutionary change could vary considerably, such variations had been ascribed simply to differences in the prevailing strength of natural selection. So by regarding speciation as the basis of change rather than its passive result, and by pointing out that species tend to remain essentially unchanged over vast spans of time, Eldredge was not only totally out of tune with received wisdom, but had turned it upside down. Making a claim in the context of a specialized paper was one thing, however; getting it widely noticed was another. This Eldredge accomplished in 1972, in an article he wrote with Harvard's Stephen Jay Gould, who had noticed the same kind of pattern in the Ice Age Bermudan land snail species he'd been studying.

The Eldredge and Gould paper, both more general and more emphatic than Eldredge's initial effort, caused an immediate stir. Perhaps its succinct and provocative title had something to do with its remarkable impact: "Punctuated Equilibria: An Alternative to Phyletic Gradualism." Starting with the proposition that preconceived frameworks affect the ways in which fossil data are viewed, Eldredge and Gould contrasted the predictions of the Synthesis with those arising from their competing theory of "punctuated equilibria," in which evolutionary change is seen as episodic. New species arise by the splitting of lineages, in a rapid but sporadic process whereby a single parent species gives rise to two daughter species. The history of each daughter species will be marked by an absence of steady change (although it's true that as it spreads out each will tend to undergo a process of geographical differentiation—mediated by adaptation to differing local habitats—which will lay the basis for species differentiation in future speciation events). The upshot is that we should not expect to find steady change through time in any local rock record (for change under the allopatric speciation model will almost invariably have happened somewhere else, in a small peripheral isolate of the parent species). Further, we should at least consider the possibility that "breaks" in the fossil record, where a species suddenly disappears to be replaced by another, are real, rather than just indications that intermediate forms are missing from that record. Acknowledging that their viewpoint was as likely as the traditional one of "phyletic gradualism" to color interpretation of the fossil record, Eldredge and Gould were nonetheless able to show that a framework of punctuated equilibria more satisfactorily explained the patterns they observed in the fossil record than did the traditional model.

Finally, Eldredge and Gould posed an apparent paradox raised by their scheme. If the evolution of species is not directional as the notion of phyletic gradualism would have it be, why do species belong to larger groups within which evolutionary trends are indeed apparent—for instance, the tendency toward a larger brain in hominid evolution? The answer to this lies in the nature of species themselves. The notion of punctuated equilibria implies that species are not simply segments of lineages that have reality only when we view them at a single instant in time. Instead, they are more like individuals, with births (at speciation) and deaths (at extinction—which can can happen to species for a variety of reasons, including being outcompeted by other species which might well be their own descendants). Thus species can be seen as playing a role on the wider ecological stage that is analogous to the one played by individual organisms in pure Darwinian theory: individuals vary in their ability to survive and give rise to offspring, and so do species as they compete with each other for ecological space. The differential survival and reproduction of species therefore account for evolutionary trends in exactly the same way the differential survival and reproduction of individuals do under more traditional constructs. Exactly why and how this winnowing of species occurs is still a matter of vigorous debate; but few would now question the importance of species as actors in their own right in the evolutionary play.

Inevitably, the notion of punctuated equilibria ran into a good deal of opposition from some paleontologists, many of whom misunderstood the point that Eldredge and Gould were trying to make. In early days it was often said, for example, that their mechanism of evolution required saltation—the quantum leaps that the Synthesis had seen off the stage. This was, of course, quite wrong; punctuated equilibria involves speciation, not saltation, and speciation is a concept that had already been squeezed into the Synthesis by Mayr and others. Another complaint was that Eldredge and Gould were against the idea of adaptation, and again this was unfair.

Certainly, there are rather few *proofs* of adaptation (parallelism—the independent acquisition in two or more species of similar specialized anatomical features—is one of them); and fine-tuning to the environment is something that, as Eldredge pointed out, it's perhaps too easy to assume when we see an organism making do in its habitat. Nonetheless, adaptation is undeniably a fact of evolutionary life, and punctuated equilibria is an adaptationist notion.

It's evident that adaptation to new conditions can take place only if physical variations that are favored by those conditions arise in a population. Goodness of adaptation being entirely relative, they will tend to do this, as will less favorable ones. New genetic and physical variants, mostly subtle, crop up in populations pretty much at random. But among widespread species the most successful of these innovations almost invariably come to be organized geographically as the species expands its range. This happens because the habitats on the periphery of the range of an expanding species are usually not typical of those in its core area, and natural selection will tend to favor innovations that are particularly useful under local circumstances. This is not simply a matter of theory, it's one of observation: as I've noted, almost any widely distributed species shows distinct regional variants whose special characteristics will generally be adaptive at some level. Of course, there's room for random factors here too, especially since for reasons of statistical sampling no isolate will be entirely typical of its species as a whole. But it's uncontested that on this local level natural selection of the Darwinian kind does take place.

The implications for species change are this: When a climatic or geographic accident separates an isolate from its larger parent population, this potential new species will already differ somewhat on average from that parent; and the distinctions existing at that time will help determine the adaptive nature of the new species. The event of speciation—a phenomenon which among mammals is very poorly understood—simply puts a permanent stamp on this divergence, by establishing reproductive isolation (which makes each new species a discrete historical entity); it need not itself have anything to do with adaptation as such. However, this process of geographical followed by reproductive isolation does have the effect of dramatically decreasing the size of the gene pool of the new species, and it is axiomatic in genetics that small gene pools are inherently more unstable than large ones. The new species will therefore be more susceptible to change than the parent—and this change, again, may be adaptive. It was probably such a speciation-related episode that Eldredge was witnessing in his sample of trilobites from that locality in upstate New York. If this new species is successful, as Eldredge's new trilobite evidently was, the population will expand, and the enlarged gene pool will become more resistant to change, setting the stage for anatomical stability. It's unlikely, however, that even after all this the new species will be very greatly different from its parent, and of course that has implications for how we recognize species in the fossil record.

We'll come back to this; suffice it to repeat here that punctuated equilibria does indeed incorporate the notion of adaptation. And while criticism of this view of the evolutionary process didn't stop with allegations of antiadaptationism, many paleontologists did readily see punctuated equilibria as the answer to problems that they were having in the interpretation of their own fossils. The upshot is that although Eldredge and Gould unleashed a controversy in evolutionary biology that reverberates still, twenty years down the line the idea has become firmly entrenched that the short-term appearances and disappearances of entire species cannot be ignored in any comprehensive account of the history of life on Earth.

The notion of punctuated equilibria has, in fact, fitted in remarkably well with a burgeoning realization of just how changeable environments have been over Earth's long biological history. For it turns out that climates and habitats have fluctuated at a pace that makes the whole idea of steady directional change over vast periods of time look rather implausible. After all, if a species finds its habitat rapidly changing, it is far more likely to migrate or to go extinct than to change on the spot; and the long-term evolutionary trends that seemed so neatly to prove directional evolution have simply persisted for too long to be explained in a seesawing world by gradual adaptation to steadily modifying environments, or by better adaptation to existing ones.

In any event, the advent of punctuated equilibria provided an entirely new perspective from which to view the fossil record, including that of Hominidae, and in 1975 Eldredge and I published an article in which we examined some of the implications of this perspective for paleoanthropology. We pointed out in particular that since the fossil record was by its nature a matter of discovery, so too, it had been thought, was evolutionary history. Effectively the belief was that if we crawled across enough rocky outcrops and discovered enough fossils, the history of the human lineage would eventually become revealed to us. This fit in well, of course, with the dictates of the Synthesis. For if change in lineages was gradual, then what we were out there discovering was equivalent to links in a chain, and all we had to do was to find enough of them to show how and where the chain ran. Deciphering human evolutionary history thus amounted to joining up extinct human species—conveniently defined for us by the "breaks" in the fossil record—on a time chart. But, we continued, if our evolutionary history was one of speciation, with more or less long-lived species replacing each other, what we actually had was a pattern of relationships in the fossil record: a pattern that couldn't be directly discovered, but that instead re-quired analysis. And clearly our current methods of analysis were faulty, whatever they were; for throughout the history of paleoanthropology new discoveries, far from clarifying the picture, had tended to make it more obscure and controversial.

If, that is, there were traditionally any clearly articulated methods of re-constructing phylogenetic history at all. When I was a graduate student, I occupied a desk in a basement storeroom of the Peabody Museum of Nat-ural History: one of the nation's great repositories of fossil vertebrates. As an aspiring paleontologist I was impatient to learn the arcana of the pa-leontologist's science. I would watch, mystified, as visiting scholars bent intently over specimens they had retrieved from collection drawers. Meas-urements that they took of skulls and bones seemed pretty straightforward (though I've since concluded that they are of less use than generally sup-posed). Intricate anatomical descriptions I understood too, for anatomy lies at the heart of what is ultimately a comparative science. But what of the other notes that these professionals so industriously made, as they pored over the fossils for hour after hour? What were they doing that allowed them to understand these documents of ancient life, beyond simply describ-ing them? In those days fundamentals such as these were barely addressed at all in the average course on vertebrate evolution, where one simply served a sort of apprenticeship in learning the facts of the history of this or that group of animals, or at least the view of them favored by the pro-fessor. Eventually I plucked up the courage to ask a distinguished scholar the crucial question: How *does* one study fossils? How *does* one understand what they tell us about the history of life? The answer? "You look at them long enough, and they speak to you."

Nowadays I realize that this response has a great deal more merit than it would appear to on the surface. Sheer intimate acquaintance with fossils, patiently acquired over long hours spent studying them, does indeed lead one to insights that go well beyond mere description and comparison. But at the time, eager as I was for a prescription telling me how to go about the conduct of the science to which I hoped to devote my career, I was frankly disappointed by this rather dusty answer. It suggested that there was no procedural Philosophers' Stone that would unlock the secret to un-derstanding the rich tapestry of life's evolution in all of its dimensions; no trick of the trade I could learn that would admit me to the inner sanctum of paleontological professionalism. All, it seemed, depended on intuition; and how does one learn intuition? How can one be taught it?

Well, despite my disappointment at the time I'd nowadays be the last person to dispute the importance of intuition in science; for there's no doubt that it stands as the very foundation of scientific creativity. And, whatever

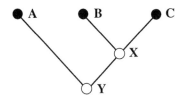

A cladogram in its simplest form.

they were, traditional methods had admittedly brought the sciences of paleontology and zoology a very long way. But they did have their limits, and as the fossil record increased by leaps and bounds their fraying edges were beginning to show. So it was a happy chance indeed that brought me in 1971 to the American Museum of Natural History, where a not-so-quiet revolution in systematics (the study of the diversity of organisms and of the relationships between them) was getting under way.

In 1950 a German entomologist named Willi Hennig published a book (in his native language) in which he articulated an approach to biological classification and the determination of evolutionary relationships that he called "phylogenetic systematics." Ernst Mayr later dubbed this approach "cladistics" (from the Greek word *clados*, meaning "branch"), because the relationships so determined were expressed in branching diagrams. Hennig's work did not attract a great deal of attention outside his homeland until 1966, when his book was translated into English. Thereafter the cladistics bandwagon began to gather momentum, as Hennig's ideas were taken up and elaborated by a group of systematists, prominent among whom were the American Museum of Natural History ichthyologists Donn Rosen and Gareth Nelson. Prior to this time, most scientists had determined relationships among organisms essentially on the basis of overall similarity between them. In paleoanthropology, for example, Wilfrid Le Gros Clark had emphasized the desirability of looking at "total morphological pattern" rather than at single characters in establishing evolutionary relationships. The problem was that it was hard for different researchers to agree on what total morphological patterns were, so that it all came down to a sort of seat-of-the-pants judgment that made competing theories of relationship hard to evaluate against each other. Hennig's great contribution was to provide a framework within which alternative hypotheses of relatedness could, in theory at least, be rigorously compared.

Forsaking the traditional representation of evolutionary relationships by lines snaking over a time chart, Hennig introduced the "cladogram," a branching diagram in which species (or larger taxa) were organized in nested sets showing their descent from a common ancestor. In the simplest case, a cladogram looks like the one opposite, which merely states that while A, B, and C are all descended from a single common ancestor, B and C share a more recent common ancestor than either does with A. How do we know this? The only way in which we can do so is to find one or more "derived" characters (evolutionary novelties) shared by B and C that are not present in A. These will presumably have been inherited from their hypothetical common ancestor X. Other characters shared by A, B, and C will have been inherited from Y, the common ancestor of all three, and simply show that all belong to the same group. Such characters will be derived for Y compared to its own ancestor; but as far as A, B, and C are concerned they are "primitive": common inheritances from their mutual ancestor that tell us nothing about relationships among the three. In drawing up this cladogram we do not need actually to know the ancestors X and Y as fossils; we simply infer that they existed from the distributions of characters (or, more properly, character *states*, or alternative forms of the same character) among A, B and C. What's more, neither time nor geographical distribution is a consideration in this process; we are concerned here only with morphology. Evolutionary relationships are thus determined exclusively on the basis of shared derived character state(s), and exactly the same basic process I've described here is followed in more complex cases involving more taxa.

How do we determine which character states within a group of taxa are primitive, and which derived? Most commonly, primitive character states are widespread within any diverse group, whereas derived ones have a more limited distribution. Moreover, if a character state is found in the next most closely related taxon outside the group, it can also be considered primitive and thus of no help in determining relationships within the group. Sometimes the process of individual development can also be informative: the fact that gill slits appear early in embryonic life among humans and other land-dwellers, for example, helps confirm that gills are a primitive character among vertebrates. A more controversial approach, much debated but potentially available for resolving problems for which no other solution seems to exist, is to assume that character states that first appear early in the known fossil record are more likely to be primitive than those first appearing later. A nagging problem with the whole thing is posed by parallelism: the independent development of similar character states. Such cases obviously tell you nothing about ancestry, while obscuring the overall picture; and perhaps the most striking result of the introduction of cladistic analysis was the demonstration that there is a lot more parallelism about than anyone had dreamed. Actually, if you have enough characters to work with, it turns out that parallelism is usually not an insuperable difficulty (computer algorithms have recently been developed that help deal with it, for instance). However, particularly if you're studying a closely knit group such as Hominidae you can never ignore it, because the more similar a pair of species is genetically, the more likely the same detailed morphology is to arise in parallel.

Whatever its difficulties, however, cladistics provides a logical procedure for the framing and testing of phylogenetic hypotheses, and it came as a breath of fresh air to one who was searching for a more satisfactory means than simple intuition of determining phylogenetic relationships. In our 1975 paper Eldredge and I included the first cladistic analysis of relationships among the hominids, and a pretty naïve effort it was too, perhaps unsurprisingly given that it was the work of a trilobite specialist and one who had up to then been interested mainly in the lower primates. It did, however, turn up some rather surprising results. The principal one was that we found it hard to squeeze the Asian *Homo erectus* in as the intermediate between *Australopithecus* and *Homo sapiens* that it was thought to be. Its long, low, flattened cranial vault, for example, is highly derived for Hominidae, but is most certainly not shared with *Homo sapiens*. What this made clear was that *Homo erectus* had been made the ancestor of *Homo sapiens* not on any compelling morphological grounds, but because it simply happened to occur at the right *time* to be that ancestor. And this was typical of the discoveries that early cladists were making at that time all over the vertebrate fossil record.

Unsurprisingly, perhaps, our article did not by itself make much of a splash. But it did usher cladistics into paleoanthropology, and since then this approach has steadily made inroads into our science. Some paleoanthropologists still reject cladistics, of course, and sometimes it's hard to avoid the impression that many of those paleoanthropologists who embraced it have more deeply absorbed its jargon (which I've avoided here) than its philosophy; nevertheless, as we'll see, the introduction of cladistics has as profoundly affected paleoanthropology as it has the other branches of vertebrate paleontology.

But the rethinking of how paleoanthropologists should go about their business didn't stop with the advent of cladograms, which actually had the effect of pointing up a more general weakness of hypothesis formulation in paleontology. In 1977 Eldredge and I suggested that one of the reasons why paleoanthropology was such a contentious business was that theories of human evolution tended to be introduced at far too complex a level. If a hypothesis is to be scientific, it has to be proposed in such a way that it can at least potentially be proven wrong: it has to be objectively testable. And, as it turns out, the only kind of testable paleontological hypothesis is the cladogram, which simply tells you which taxa are most closely related to which others. It says nothing about the nature of the relationships involved. Such relationships may be of two kinds: that between an ancestor and its direct descendant, and that between two "sister" taxa descended from the same ancestor. If you add ancestry and descent (which usually also involves time) to your cladogram, you get what's called an "evolutionary tree." Johanson and White's hypothesis about *Australopithecus afarensis* was a formulation of this kind. But since it's not actually possible to *prove* ancestry (or, in some cases, to disprove it), trees are not only more complex statements than the cladogram you started with; they're also not testable. And because you can derive a number of different trees from a single cladogram, this obviously leaves the door open for endless arguments.

Yet more complex than the tree is the "scenario." This is what you get when you add the really interesting stuff to to the information already present in the tree. This added information includes everything you know about adaptation, ecology, behavior, and so forth, and it's certainly what makes the past come alive. But it does mean that the average scenario is a highly complex mishmash in which considerations of relationship, ancestry, time, ecology, adaptation, and a host of other things, are all inextricably intertwined, tending to feed back into each other. When you're out there selling such complicated narratives, normal scientific testability just isn't an issue: how many of your colleagues or others buy your story depends principally on how convincing or forceful a storyteller you are—and on how willing your audience is to believe the kind of thing you are saying (which brings us back to the importance of people's expectations). Of course, this is no reason to abandon scenarios and restrict paleoanthropology to the intrinsically more limited and less interesting statements represented by cladograms and trees. But besides laying out these different levels of analysis, what Eldredge and I were at pains to emphasize was that whenever a broad interpretation of an episode of human evolution is offered, the analysis should proceed from the simple to the complex: you should start with a cladogram, advance to a clearly justified tree, and only then go on to the scenario. This way, it will be clear to everyone not only what the basic testable element is in each case, but also how the more complex hypotheses have been arrived at. Thus a basis at least will exist for discussion and comparison of those more complex hypotheses. The problem in anthropology as we saw it, and the cause of endless confusion and misunderstanding, was that people were diving in at the deep end and starting with scenarios.

Well, old habits die hard, but progress has been made. Although the cladistic approach to phylogenetic reconstruction has not been adopted by all paleoanthropologists, cladograms have become a fact of paleoanthropological life and can hardly be totally ignored by anyone offering a scenario. Similarly, although the gradualistic view of human evolution continues to predominate in many quarters, there is an increasing awareness of the need to define and recognize accurately the species that have existed in our evolutionary past. Exactly how this should be done remains a contentious issue, which we'll look at in a later chapter; meanwhile, let's return to the fossil record itself.

13

Eurasia and Africa: Odds and Ends

As we've seen, during the 1960s and 1970s paleoanthropological attention was focused largely upon Africa, and upon the earlier phases of human history. But these weren't quite the only games in town. Europe and Asia were also yielding up new fossils, and new dates were being established for old ones. On the Asian front, the most remarkable discovery of the period was a fairly complete cranium of *Homo erectus* (known as Sangiran 17), found by the Indonesian paleontologist Sastrohamidjojo Sartono in 1969. This specimen is larger than the skullcaps found before the war (its brain volume is just over 1,000 ml) and considerably more robust. Its thicker brow ridges, sharper cresting across the back of the braincase, and generally heavier build suggest that if all of these specimens belong to the same species, the new skull is that of a male while the earlier Javan skullcaps are females. Sangiran 17 is particularly interesting because it preserves much of the structure of the face. This is massively built, with huge cheekbones that flare toward the side, and it projects somewhat.

Several other more fragmentary hominid fossils were also found by Sartono and the paleoanthropologist Teuku Jacob at sites in central Java, and the whole known assemblage has recently been reviewed by Philip Rightmire of the State University of New York at Binghamton. Rightmire finds no major difficulty in fitting all the Javan hominid fossils into the species defined by Dubois's original Trinil skullcap; but on this, as on virtually all paleoanthropological subjects, the last word has yet to be written. Continuing efforts to date *Homo erectus* from Java have revealed substantial problems with the geology and biostratigraphy of the area; at present the best estimate is that most of the *Homo erectus* fossils from the island are between about 1 myr and 700 kyr old. The Ngandong (Solo) specimens are usually

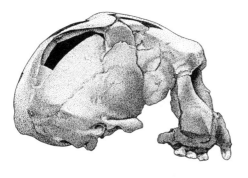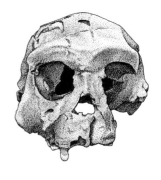

Side and front views of the Sangiran 17 skull from the Kabuh Beds, Sangiran, Java. Scales are 1 cm. DM.

thought to be very substantially younger than this range, but their provenance is so poorly documented that a date of 500 kyr (or even more) is not entirely out of the question.

Important human fossils also continued to emerge from China. Zhoukoudian produced another skullcap in 1966; this proved to articulate with a cast of one of the temporal bones which was lost with the other fossils from the site in 1941, and it fits well among the prewar skullcaps. Using a combination of dating techniques, Chinese researchers were able to estimate a relatively late age span for the Sinanthropus deposits of Zhoukoudian,

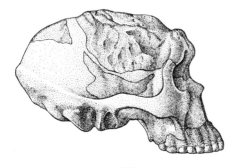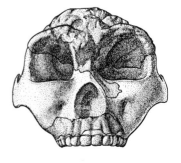

Side and front views of the reconstructed cranium from Lantian, China. Scales are 1 cm. DS.

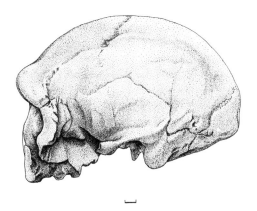
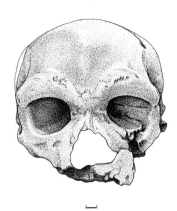

Side and front views of the damaged and distorted cranium from Dali, China. Scales are 1 cm. DM.

from about 400 to 225 kyr ago. From Lantian County of Shaanxi Province came, in 1963 and 1964, a mandible and skullcap which may be about 700 kyr and 1 myr old, respectively; these thus fall in the time range of the Javan remains and despite rather poor preservation do not look strikingly different from them. Another skullcap of the same general form, plus some isolated teeth, came in 1980 from Hexian County, Anhui Province; this site too is poorly dated but may be about 200–400 kyr old. More recently, the site of Jinniushan in Liaoning Province produced unusually complete ancient human remains consisting of a cranium with a reported brain capacity of almost 1,400 ml, plus a variety of postcranial bones. These fossils apparently date from about 200-300 kyr ago, and although they have usually been referred to as *Homo erectus* they may in fact be a little more modern. Unfortunately, as with most postwar Chinese finds, they have yet to be described in useful detail.

A human cranium from China that is definitely more modern than *Homo erectus* was found in 1978 in Dali County, Shaanxi Province. This specimen has a brain capacity of about 1,200 ml and sports large brow ridges; but the face, unfortunately a bit crushed, is rather lightly built and doesn't project much. Although adequate description of the Dali cranium is slow in seeing the light of day, this specimen seems to fit among the "intermediate" group, exemplified by the Kabwe cranium from Zambia, that lies, in time at least, between most *Homo erectus* and *Homo sapiens*. This is reflected in the way in which Chinese investigators have classified it: it was initially described by Y. Wang and collaborators as *Homo erectus*, while later Wu

Xinzhi of the Institute of Vertebrate Paleontology and Paleoanthropology in Beijing placed it in a separate subspecies of our own species, as *Homo sapiens daliensis*. Dating is a problem once more; the specimen is probably about 150 kyr old, though it may be a little more ancient than that. This would make it about the same age as a skullcap found in 1958 at Maba, Guangdong Province, which is lightly built but has a distinct brow ridge; it is generally regarded in China as an early *Homo sapiens* skull with affinities to modern Asians, while retaining some features of *Homo erectus*. A similar claim has also been made for a couple of crania, allegedly about 350 kyr old, that were found in 1989–1990 at a site in Hubei Province.

Such allocations reflect the long-prevailing wisdom that any Middle Pleistocene fossil human must belong either to *Homo sapiens* or to *Homo erectus*, and this is a mindset that afflicts European paleoanthropologists quite as much as it does their Chinese counterparts. In 1960, for example, a cranium was found in a cave at Petralona, in northeastern Greece. This is a marvellously preserved specimen that has only recently been totally cleaned of the calcite coating that had covered and protected it. Unfortunately, the calcite matrix that was removed was not kept, for it might have been used for dating using a new technique that I'll explain later; as it is, the specimen may be anything from about 250 to about 600 kyr old, with the consensus at about 400 kyr. The Petralona fossil itself is clearly neither *Homo erectus* nor *Homo sapiens*: it had a relatively large brain of about 1,200 ml in volume, but the cranium, while better inflated than that of *Homo erectus*, is longish, with a clearly retreating forehead behind large brow ridges, and is clearly angled at the back. The face is large and projects distinctly, especially in the midline. Early reports described this specimen as belonging variously to *Homo erectus* and *Homo sapiens*, but most opinion veered quite quickly towards an "archaic" form of the latter, clearly distinct from both modern humans and Neanderthals. Recent studies have, however, detected some Neanderthal-like features in its face.

Quite similar are several hominid fossils excavated by Henry de Lumley and his colleagues at the cave of Arago (Tautavel)in the foothills of the French Pyrenees, close to Perpignan and the Spanish border. Work beginning in 1964 at this remarkable site revealed a stratified succession of living floors, with an abundance of animal bones and rather crude stone tools (mostly flakes of one kind or another; rather few handaxes). The age of the deposits has proven difficult to determine, but most would now reckon that the human fossils are about 400 kyr old, give or take 100 kyr or so. Among the first of them to show up were a couple of partial lower jaws that are both robust and lack chins, but which differ noticeably in size. This difference is generally ascribed to sexual dimorphism. The most interesting

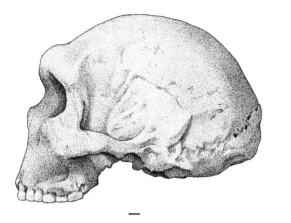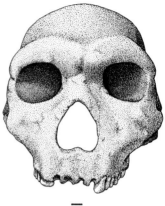

Side and front views of the cranium from the cave of Petralona, northern Greece. Scales are 1 cm. DM.

specimen, found in 1971, consists of a well-preserved if somewhat distorted face, together with an apparently associated parietal bone. These two elements have been used to produce a reconstruction of the cranium, which has an estimated volume of 1,100–1,200 ml. Overall, the reconstructed skull is generally quite Petralona-like, with a modestly inflated cranial vault that recedes behind marked brow ridges. The face projects modestly and is rather lightly built; this has led some to suggest that the cranium is that of a female, although its describers preferred a male designation. In early days the Arago hominid was often described for convenience as a "pre-Neanderthal", more for reasons of dating than of morphology; recent studies have actually detected some Neanderthal-like traits, particularly in the lower jaw. The closest comparison, however, is clearly with Petralona in Europe and with the more robust Kabwe and Bodo specimens in Africa.

The de Lumley group has backed the allocation of the Arago hominids to an advanced form of *Homo erectus*, and this usage has generally been followed by continental European paleoanthropologists (because clearly these and similar fossils are not *Homo sapiens*). English-speaking scientists, on the other hand, have mostly preferred to regard them as belonging to "archaic *Homo sapiens*" (because, equally clearly, they are not *Homo erectus*). Here we have a perfect example of how expectation has colored interpretation of the fossil record. Because received wisdom tells us that *Homo erectus* evolved gradually into *Homo sapiens* (and because it admits no named intermediate stages), we should expect to find forms that might be classified either way. And since Arago and its like are at least roughly intermediate between these species both in time and, at least as important, in brain size (an attribute which appeals powerfully both because it is so easy to quantify and because it somehow expresses the essence of humanness), we not only expect to find them by their nature impossible to classify, but we don't bother to examine their morphology very closely. If we did, however, we might reach a very different conclusion about their relationships: a point to which we will return.

To sum up, then, there was a lot of very vocal and ultimately fruitless discussion of these various new finds from Asia and Europe. But aside from eliciting differences of opinion on what to call them, they initially did rather little to stimulate any rethinking of the course of human evolution. This, again, can reasonably be ascribed to the gradualist mindset that prevailed when they were found: the new fossils simply needed to be slotted somehow—anyhow—into a preexisting framework, after which what they actually looked like could safely be ignored. Little wonder that more interesting things were emerging from archaeological field investigations than from the paleoanthropological laboratories.

Before turning his attention to Arago, for example, Henry de Lumley undertook a salvage excavation of a site known as Terra Amata, in the southern French town of Nice. This site, more securely dated than Arago to about 400 kyr ago, is interesting from several points of view. Among other things it contains, in the form of hearths, what may be the earliest evidence in Europe of the domestication of fire (although it's possible that the Spanish sites of Torralba and Ambrona, which may be as old or even a little older, should take the laurels here). More important, though, Terra Amata apparently represented a seasonal hunting camp whose inhabitants built shelters of saplings placed into the ground in ovals and brought together at the top. If this interpretation is correct, Terra Amata provides the earliest evidence from anywhere of such activity.

Throughout the postwar period the faunal succession in Europe was being refined and calibrated, and the earliest evidence for human occupation in Europe was being pushed back. Schoetensack's Mauer jaw was faunally dated to over half a million years ago, and remains the earliest fossil evidence for ancient human presence in western Europe. Archaeological indications appear to go back further, though. Aside from a claim, still disputed, of crude 2 myr old tools from a site in central France, several sites excavated during this time offer evidence for human activity in the western areas of Europe in the period following about 1 myr ago. Significant among these are the localities of Soleilhac and Le Vallonet in France,

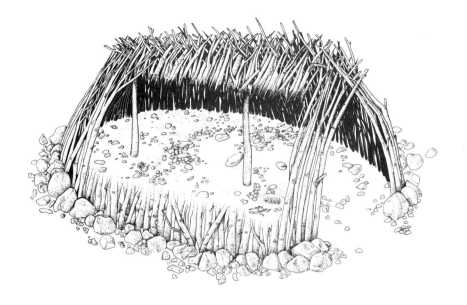

Artist's view of one of the hutlike structures at Terra Amata, France, with side cut away to show a hearth and interior debris. Drawing by DS after a concept by Henry de Lumley.

and Isernia La Pineta, in Italy, all of which contain simple flake tools and appear to be over 700 kyr old. A very recent find of a hominid mandible at Dmanisi, in ex-Soviet Georgia, however, places the entry of humans into Eurasia much further back than anything documented from the west; the specimen may be as much as 1.6 myr old, and is no younger than 900 kyr.

Further south, excavations during the early 1960s at the site of 'Ubeidiya, in Israel, produced Acheulean artifacts that are pretty firmly dated to about 1 myr ago or even earlier. This is good proof that by this time handaxe makers had managed to leave Africa, where these bifacially fashioned tools were first produced some 1.5 myr ago. It's thus maybe a little surprising that the earliest European archaeological sites are bereft of handaxes; but it seems to be a general phenomenon that these tools get rarer the further away from Africa you go. Indeed, as far back as the 1940s Hallam Movius noticed that Paleolithic tool assemblages from India east to the Pacific rim tended to lack handaxes entirely, consisting instead purely of chopper and flake tools. The significance of the "Movius line" between the handaxe-making cultures to the west and the nonhandaxe-makers to the east has long been argued over. The latest wrinkle is that it coincides more or less with the western distribution limit of bamboo: a versatile material that may well have served as an excellent substitute for stone. Well, maybe.

In Europe, handaxes were being produced occasionally by Arago times, and extraordinary abundances of these implements were reported by early workers at not much younger sites in France's Somme River valley, where Boucher de Perthes had made his original observations. Nowadays, though, it seems likely that this apparent great density of handaxes (which recalls a pattern widely seen in Africa) is in fact the result of selective collecting by nineteenth-century workers. More recent controlled excavations at European sites of Acheulean age have tended to show not only that a rather limited range of tools was being produced, but also that handaxes were not necessarily numerous among them—sometimes, indeed, they were rare or even lacking altogether. It's still unclear whether those sites not containing bifaces represent a separate cultural tradition from the biface sites, or whether they are simply a product of local circumstances or even sampling error.

No less obscure is the nature and timing of the "transition" between the Lower Paleolithic period (in Africa, the Early Stone Age) to which the Acheulean belongs, and the Middle Paleolithic (Middle Stone Age) period which followed it. At some point around 200-250 kyr ago, in both Africa and Europe, Acheulean cultures based on bashing a core into a particular shape began to be elbowed aside by stoneworking industries based on "prepared core" technology. This involved shaping a stone core in such a way that a flake of predetermined size and form could be struck from it at a single blow. With relatively little fuss this flake was then made into the finished tool. Such tools were generally scrapers of one kind or another, but quite often they assumed the form of handaxes. The prepared-core technique had several advantages over earlier ways of stoneworking: one was a greater economy of material (for several flakes were often obtained from a single core), and another was the ability to control the shape of the flake; but probably the greatest advantage was that it produced a tool with a long continuous cutting edge.

Dating of Middle Paleolithic sites is generally rather poor, especially since they lie beyond the range of radiocarbon. For this reason alone it's hard to detect what was actually going on as handaxes began to decline and prepared core tools became increasingly common. There's no doubt, however, that this was not a smooth, gradual process. What's more, prior to the Neanderthals there is little direct evidence of the hominids who were involved in this technological changeover. It's only with the Neanderthals that the European human fossil record begins to pick up in a significant way, and by their time the Middle Paleolithic was firmly established; indeed, the Neanderthals' "Mousterian" culture was the apogee of Middle Paleolithic stoneworking.

Until recently, the end of the Middle Paleolithic posed as many problems as its beginning. The Mousterian is followed in Europe by the Aurignacian culture, which is undoubtedly the product of early modern humans. However, at certain sites the early Aurignacian is overlapped by an industry known as the Châtelperronian, an industry having technical features in common not just with the Mousterian but also with the Aurignacian and later "Upper Paleolithic" cultures. Notably, while flake tools were still important, about half the tools in the Châtelperronian kit were made on "blades": narrow flakes more than twice as long as wide, which were retouched to produce a varied assortment of different tools. Blades were the hallmark of Upper Paleolithic stone technology in Europe (although, interestingly, they never caught on quite as much in Africa, where they were first produced as long ago as 100 kyr).

Who made the Châtelperronian? Neanderthals or moderns? Was the Châtelperronian the last phase of the Middle Paleolithic, or did it herald the Upper Paleolithic? As long as there was no firm archaeological association between this industry and human fossils, it was impossible to be sure. But in 1979 a find was made of a Neanderthal burial at a place called St-Césaire, in western France. The layer containing this fossil was was late in time; even though its initial estimated date of 32 kyr was later revised backwards to 36 kyr, this is still the most recent Neanderthal fossil that is both substantial and well-dated. What's more, it falls squarely in the middle of a time gap previously unfilled by western European human fossils of any kind. More important yet, the associated tool kit was Châtelperronian. For most Paleolithic archaeologists, the St Césaire find solved the conundrum of the Châtelperronian: yes, it was the last gasp of the Neanderthals in Atlantic Europe (or was at least part of it). But why did it have such strong Upper Paleolithic features, in the very region where the late occurrence of Neanderthals demonstrated beyond doubt that they had not evolved into modern humans? One intriguing suggestion is that the western European Neanderthals learned by observation to make Upper Paleolithic-style blade tools, as early modern humans began to trickle into their territory at some time after about 40 kyr ago, bringing this new technology with them.

But even if this interpretation is correct, other evidence suggests as messy a picture for the departure of the Neanderthals as for the arrival of the Middle Paleolithic. At Portugal's Figueira Brava Cave, for example, fragmentary Neanderthal remains have recently been found in levels dated to about 31 kyr ago. These fossils, 5 kyr younger than those of St-Césaire, were found in an "evolved Mousterian" archaeological context and appear

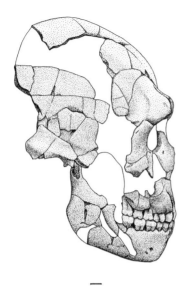
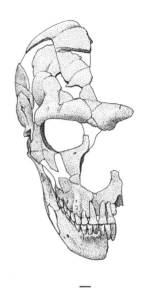

Side and front views of the partial Neanderthal skull from St-Césaire, western France. Scales are 1 cm. DM.

to be good evidence for late but technologically pristine Neanderthal survival on the Iberian peninsula. Ironically, it is just this area which is beginning to yield evidence for the very early arrival of modern humans in Atlantic Europe: looking upon Iberia as the last redoubt of the Neanderthals against invading hordes of moderns arriving from the east begins to seem clearly oversimplified. As usual, the plot thickens.

The beginning of Neanderthal tenure in western Europe is also being pushed backward by new finds and new dates. In the late 1970s the back of a cranium was found at Biache-St-Vaast in northeastern France. The associated industry has been described as Mousterian, and the age of the site appears to be in excess of 150 kyr. The fossil shows "bunning" (a rounded backward protrusion) of the occiput and other features typical of the Neanderthals; despite its incomplete nature there seems to be no doubt that it represents a fully fledged Neanderthal individual. In addition, recent studies of the badly fragmented Ehringsdorf material initially described by Franz Weidenreich have indicated that it possesses a number of characteristically Neanderthal features. Further, dating carried out in 1982 using new techniques suggests that these fossils may be around 200 kyr old, though a more realistic assessment might indicate a broader possible time range, between about 110 and 200 kyr. By the early 1980s, then, it was possible

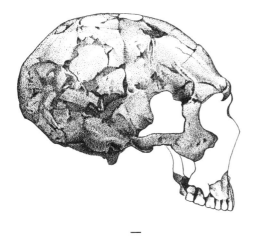
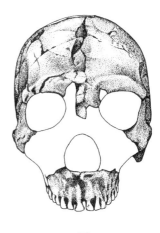

—   —

Side and front views of the cranium from the cave of Amud, Israel. Scales are 1 cm.   DM.

to say that the Neanderthals were around in western Europe as a clearly recognizable group at least 150 kyr ago, and quite possibly well before that. Just how much before that was thrown into doubt in 1993 when Juan-Luis Arsuaga and colleagues from the Universidad Complutense in Madrid reported the discovery of three quite complete hominid skulls from the 300 kyr old site of Sima de los Huesos in the Atapuerca Mountains of Spain. These specimens quite closely resemble the Petralona-Arago group, but they also exhibit a number of Neanderthal features. At this writing, the paleoanthropological community has yet to digest these significant finds.

How about Neanderthals farther east? In 1961 excavations by a Japanese team at the Israeli cave of Amud revealed in a Mousterian layer a fairly complete skeleton of a young adult male Neanderthal, with the extremely large cranial capacity of 1,740 ml (which may have to do with the fact that this is also the tallest Neanderthal yet discovered, with a stature of about five feet ten inches). Dating estimates are converging on an age of about 60 kyr for this skeleton (and an infant skeleton very recently discovered). Both are clearly Neanderthal, and the adult is quite similar to the Shanidar specimens, confirming that even quite late Near Eastern Neanderthals failed to show the exaggerated characteristics of their kind exhibited by their "classic" contemporaries in western Europe. If the Amud individual was unusually tall, then another Neanderthal from the not too far distant and approximately contemporaneous Israeli cave site of Kebara was unusually

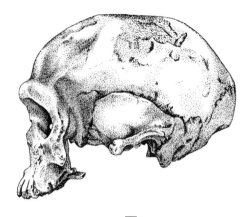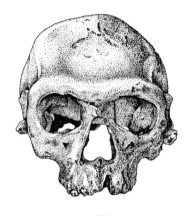

Side and front views of the Jebel Irhoud 1 cranium from Morocco. Scales are 1 cm. DM.

massive: extremely strongly built, and with the largest lower jaw yet known for a Neanderthal. The Kebara fossil consists of most of the skeleton of a male individual who was apparently deliberately buried, but it lacks the cranium even though the mandible is present. Very unusually, the hyoid (throat) bone is preserved, but that is a story we'll address later. The Kebara individual is the most recent Neanderthal we know of from the Near East, dating from about 50 kyr ago, and was found in a Mousterian archaeological context; however, there is clear evidence of hearths at this site, something at best unusual in European Neanderthal localities.

The 1960s saw important new finds elsewhere in the Mediterranean Basin, where during that decade the Moroccan site of Jebel Irhoud produced two crania, a juvenile mandible, and a couple of postcranial fragments. At the time the date of these specimens was exceptionally poorly documented, but their anatomy is extremely interesting. The more complete specimen has a lowish braincase and a rather big face with fairly noticeable brow ridges, but otherwise is rather modern in appearance. The less complete one has a more modern-looking front, but the rear of the braincase has been viewed as rather primitive. First reported as Neanderthal remains, these specimens are nowadays viewed as vaguely archaic moderns. Africa has still not produced any fossils that can plausibly be viewed as Neanderthal. The stone tools putatively associated with the Jebel Irhoud fossils are, however, generally of Mousterian type, as are those made by other early modern or modernish humans in the circum-Mediterranean area, such as those from Skhūl and Jebel Qafzeh.

Around the time that the last discoveries were being made at Jebel Ir-houd, more controlled excavations at the caves of the Klasies River Mouth were producing very suggestive evidence of early modern human occupation near the southern tip of Africa. The only human fossils found at the site are extremely fragmentary; they are nonetheless very clearly modern in form, although some are more robust than others. Their archaeological context, however, is thoroughly Middle Stone Age. At Klasies, as in certain other southern African sites, there is an odd intrusion into the Middle Stone Age sequence known as the Howieson's Poort industry. Middle Stone Age tools consist mostly of flakes struck from prepared cores, but the Howieson's Poort culture produced lots of blades and tiny tools, thought to have been hafted, which are often referred to as "microliths." This anticipates developments typical of the Late Stone Age in Africa (and which only cropped up in the latest phases of the Upper Paleolithic in Europe); but, occurring only over a short period about 70 kyr ago, the Howieson's Poort levels are overlain at Klasies by more Middle Stone Age deposits. Most of the human fossils from Klasies come from pre-Howieson's Poort sediments, and recent datings by various methods have suggested that the most ancient of them may be about 120 kyr old. Very intriguingly, Hilary Deacon of the University of Stellenbosch has suggested that these fragmentary Middle Stone Age humans from Klasies represent leftovers from cannibal feasts; he also finds evidence at the site for the kind of organization of the living space that is usually associated only with behaviorally modern humans.

Between 1940 and 1974 the site of Border Cave, between Swaziland and South Africa, produced several hominid specimens. All of them are without doubt anatomically modern; the problem lies with their dating, for the earliest discoveries were not made under conditions of controlled excavation. This unfortunately goes for the best specimen, an astonishingly modern-looking adult cranium that may be as much as 90 kyr old but which might have been part of an intrusive burial from a later period. Its significance, then, remains moot, although it's not unlikely that it does provide supporting evidence for very early modern human occupation of southern Africa. Nonetheless, even if both the Klasies and Border Cave dates are reliable, the South African picture of Late Pleistocene human evolution is still not all that clear. For while a partial cranium found in 1932 at Florisbad, near Bloemfontein, is approximately equivalent in age to the Klasies fossils, it is distinctly archaic in form. In some respects it resembles the Ngaloba cranium found by Mary Leakey's team in the later deposits at Laetoli, which is dated at 130–150 kyr.

A couple of further African finds should be mentioned here. In 1973

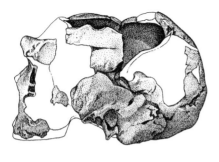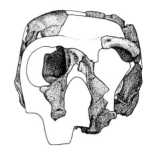

—   —

Side and front views of the reconstructed partial cranium from Lake Ndutu, Tanzania. Scales are 1 cm.   DM.

Amini Mturi, of the Tanzanian Department of Antiquities, found a partial and badly fragmented human cranium on the shores of Lake Ndutu, at the western end of Olduvai Gorge. A best estimate would give this specimen an age of about 300–400 kyr. As painstakingly reconstructed by Ron Clarke, the Ndutu fossil looks more modern, or at least more gracile, than the crania from Kabwe and Bodo, both of which may be of around the same age. There is a distinct if thinnish brow ridge, and the bones of the skull roof are rather thick; but the cranium is quite short from back to front, and it may have been relatively high. Clarke himself has drawn attention to similarities to the rather younger German cranium from Steinheim. It's also possible that this fossil belonged to the same population represented by the approximately contemporaneous and rather small-brained (about 950 ml) specimen from Salé, in Morocco; interpretation of the latter, however, is hampered by what seems to be an abnormally developed skull rear.

As this far from exhaustive list of individual specimens shows, there was throughout the 1960s and 1970s a steady background ticking of human fossil finds quite apart from the Leakey and Johanson discoveries that garnered most of the public attention. And every one of the fossils discussed in this chapter has its own particular importance for the reconstruction of the complex human past. But, to repeat, none of them in itself forced any major reevaluation of the accepted scheme of human evolution. At least in part, this was because paleoanthropologists were still pretty much stuck with the traditional way of doing things. Their role, as I suggested earlier, was still largely seen as a sort of service industry: stuffing new fossils into a convenient place in an established scheme. And it usually took a lot more than one fossil to make them think about substantially rearranging the larger picture.

Another reason, however, was quite simply that over these two decades new discoveries were coming in so thick and fast that paleoanthropologists didn't have the time or the perspective to digest them adequately. It's quite possible that the 1980s were an altogether more contemplative time in the science of human evolution because the pace of new discoveries slowed down somewhat. But although fewer new human fossils came to light, the impact of those that were found was nonetheless extraordinary. We'll look at those new finds in the next chapter.

14

Turkana and Olduvai—Again

By the time the 1980s came around few had any doubts that the human lineage had originated in Africa. And, quite possibly, fewer still doubted that *Homo erectus* stood square in the center of the road from *Australopithecus* to *Homo sapiens*. Yet *Homo erectus*, despite the Koobi Fora group's recent discoveries, remained steadfastly identified with Asia. Until, that is, the forays by Richard Leakey's team into the hot, harsh badlands to the west of Lake Turkana began to pay off. In August 1984 a member of Leakey's team, the veteran fossil finder Kamoya Kimeu, found the first tiny fragment of a hominid skull at a site next to a dry stream bed extravagantly known as the Nariokotome River. Within a month or so, the group had recovered the greater part of the skeleton of a young male whose teeth had erupted about to the stage characteristic of a modern eleven- or twelve-year-old. This skeleton, dated to about 1.6 myr ago, and immediately assigned by its discoverers to *Homo erectus*, was yet more complete than that of Lucy, and incomparably more so than any other skeleton known from before the time of the Neanderthals. It was a unique find, and an astonishing one.

For this young male totally contradicted the stereotype of *Homo erectus* as squat, heavy-boned, and powerfully muscled—a stereotype that, in the absence of substantial postcranial fossils, had endured ever since Dubois's time. Instead, the "Turkana Boy" (technically known as KNM-WT 15000) was both tall (about five feet four inches at the time of his death, but it's estimated that had he lived to adulthood he would have achieved six feet) and slender. Indeed, according to Alan Walker who led the team that studied his skeleton, he was built much like the people who live around Lake Turkana today—people whose elongated limbs and bodies are good at shedding the heat load mercilessly imposed by the sun in this arid tropic region. The prime importance of the Turkana Boy is that he represented the earliest kind of human we know of whose general body proportions matched those of living people. Not that he was modern in all respects. The upper part of the canal through which the spinal cord runs is narrow, perhaps suggesting that nervous signals to the thorax were limited. It's suggested that this may even indicate a less precise command of voluntary respiration, which might reflect a limited ability to communicate using complex and precisely controlled sounds. He was not barrel-chested, as we are, but had a somewhat upwardly tapering thorax (though less so than in Lucy or in an ape). His shoulder joints would thus have been closer to the midline of the body than ours are, and this is particularly interesting because it is helpful to be built like this when you are suspending yourself from tree branches, but it's not at all favorable in a striding biped that swings its arms for balance. The femoral heads are large, like ours, but the necks attaching them to the shafts are long, like an australopithecine's. This latter characteristic might be related to a rather narrow pelvic canal; the combination of the two factors might have enhanced stabilization of the hip, while, if the male can serve as an analogy for the female, it restricted the maximum size of the newborn's head.

This probably wasn't a problem. The skull of the Turkana Boy shows that his brain was not large; even at adult size, it would probably have been no larger than that of the slightly older ER 3733. His face, however, was more massively built and more projecting than the latter's; since 3733 is thought to be female, this difference may be due to relatively strong sexual dimorphism in the species. Interestingly, Holly Smith of the University of Michigan has shown that tooth development in the Boy had been rapid compared to modern humans, even if slowed down compared to apes and australopithecines. Thus it appears that the Turkana Boy may well have died after only about nine years of life, rather than after the eleven to twelve years a modern human takes to achieve a comparable eruption of the teeth. Below the neck, however, there's no denying that we have in the WT 15000 skeleton the first evidence of an essentially modern human anatomy. Nobody is going to dispute the Turkana Boy's abilities as an upright biped. So, once again, locomotor innovation had led the way among the evolutionary changes that ultimately led to the modern human condition. It certainly preceded any major technological improvements. For the Turkana Boy lived a full hundred thousand years before the inhabitants of the Turkana region began to make handaxes. His contemporaries made stone implements not very different from those that their predecessors had made almost a million years earlier.

Leakey and his collaborators were pretty categorical about ascribing their new skeleton and comparable Turkana fossils to the long familiar species

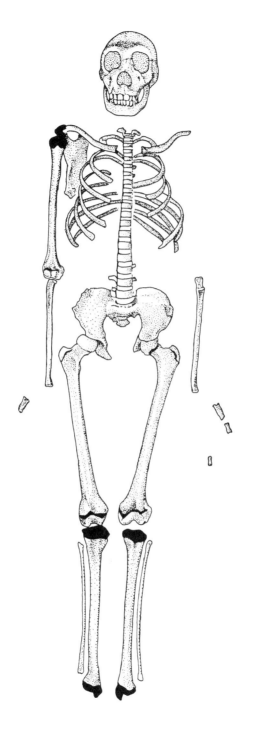

The "Turkana Boy" skeleton (KNM-WT 15000) from Nariokotome, West Turkana, Kenya.   DS.

Homo erectus. Allusions to the Zhoukoudian hominids, in particular, frequently cropped up in their discussion of these materials. But even as they were making their great discovery, doubt was beginning to be cast on whether the allocation to *Homo erectus* was was actually appropriate. At a 1984 meeting in Germany held to honor the memory of Ralph von Koenigswald (who had died in 1982), some of the participants questioned whether "early African" and "later Asian" *Homo erectus* were indeed the same thing. Peter Andrews, who summarized the proceedings, put the matter well. The problem in defining *Homo erectus*, he said, lay in the fact that it was "viewed at present as a grade of human evolution intermediate between the small brained early Pleistocene hominids and the large brained *Homo sapiens*." In short, brain size was the key, and other characters were ignored. Yet brain size could not by itself be considered proof of membership of a particular species. Further, Andrews emphasized the point made by his London colleague Chris Stringer that many of the fossils lumped together into *Homo erectus* were linked simply by primitive characters inherited from a remote ancestry, rather than by derived ones that might indicate a special relationship. He pointed out that the Asian fossils were characterized by a suite of derived traits that were not present in the African forms, and he implied that the two populations actually belonged to distinct species. Examining the connotations of all this, Andrews found that the simplest scenario of *Homo sapiens* evolution "bypassed *erectus* in Asia," thus echoing the conclusion that Niles Eldredge and I had reached back in 1975. Cladistics was beginning to bite.

So were other new fossils. In 1985 Don Johanson and his collaborators received permission from the Tanzanian authorities to reopen fieldwork at Olduvai Gorge, from which Mary Leakey had retired several years earlier. In July 1986 they found the fragmentary remains of a hominid skeleton that they called OH 62. And fragmentary it was indeed. The skeleton had been eroding out of the gorge deposits for a very long time—perhaps even for centuries, it was estimated—and was shattered into hundreds of tiny fragments, many of which were retrieved only by dint of laborious sieving of the sediments. But even with a total of some three hundred shattered bits not much of the skeleton was preserved; only the upper jaw (with some teeth) and some other skull fragments, most of the right arm and parts of both legs remained. Although this was not what the team had hoped for in the excitement of the initial discovery, it was enough to draw some pretty startling conclusions about *Homo habilis*, the species to which it was assigned on the basis of resemblances in its teeth and palate. This assignment fit pretty well with its age and provenance: the site of discovery lay close to the bottom of the Gorge, not far from the famous Zinjanthropus site, and fell between two tuffs dated at 1.85 and 1.75 myr old. This made it almost exactly the same age as the original *Homo habilis* type specimens—and a bare couple of hundred thousand years older than the Turkana Boy.

The rather close correspondence in time between these two fossil individuals, both allocated to *Homo*, made the new Olduvai find seem especially remarkable. For if the Turkana Boy was tall and strikingly modern in his below-the-neck anatomy, OH 62 was quite the reverse. In life OH 62 had probably stood even shorter than Lucy; and it (we don't know what sex he or she was) had probably walked in a similar way, too, because its limb proportions seem if anything to have been even more archaic than Lucy's. In particular, OH 62 had had long, powerful arms, and Sigrid Hartwig-Scherer and Bob Martin of the University of Zürich have recently shown that by a variety of measures its limb bones had closer resemblances to those of apes than did Lucy's. All of this came as a surprise, to say the least, to a profession dominated by the gradualist mindset. Expectation had been that the body skeleton of *Homo habilis*, when found, would at least be intermediate in its morphology between those of *Australopithecus* and *Homo erectus*; this expectation was so powerful that, for example, some 2 myr old isolated limb bones from East Turkana had frequently been ascribed to *Homo habilis* purely on the basis of their rather modern appearance.

Of course, at least part of the problem lies with the unsatisfactory catchall nature of *Homo habilis*. In describing their new fossil, Johanson, Tim White and colleagues noted that the original Olduvai Bed I *Homo habilis* had met with the objection that distinction from *Australopithecus africanus* was unwarranted. And they also claimed that new discoveries (including ER 1470 and 1813 and a new skull called Stw 53 that Alun Hughes had found in Member 5 of Sterkfontein in 1976—and which was found to be a particularly good match for OH 62) had disposed of such objections. Yet in their very next sentence they noted that the new fossils had given rise to attempts to break up this larger sample into more than one species. For, despite the convenience it represented, rumblings were beginning to be heard by the mid-1980s that the newly enlarged *Homo habilis* might be a somewhat ill-assorted mishmash of fossils belonging to more than one kind of early hominid. Thus at the "Ancestors" meeting held in 1984 at the American Museum of Natural History to inaugurate the first large public exhibition of original hominid fossils, the anatomist Bernard Wood had stated pretty flatly that there were at least three "non-australopithecine taxa" in the east African early Pleistocene. Similarly, in 1986 Chris Stringer published a paper called "The Credibility of *Homo habilis*" in which he found evidence for "at least three Plio-Pleistocene species of 'early *Homo*'" in eastern Africa (effectively, Turkana and Olduvai).

Stringer also noted that if you ignored cranial capacity, fossils such as ER 1470 and OH 24 looked more australopithecine than was generally ac-

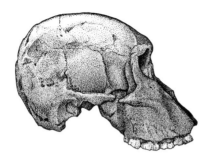
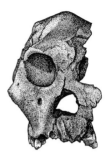

Side and front views of the Sts 71 cranium from Sterkfontein Member 4, South Africa. One of the more "robust" specimens from the site. Scales are 1 cm. DM.

knowledged. Which brings us to the second part of the problem posed by *Homo habilis*. For, even if the gracile hominid fossils from Olduvai and Turkana are indeed excludable from *Australopithecus* (which is, I think, at least marginally justifiable), are they properly assignable to *Homo* (which in my view is much more debatable)? Here again we are heirs to an unfortunate legacy of the Synthesis. We've seen that lumping of species is viewed by Ernst Mayr's many followers as antitypological, and therefore as by definition a Good Thing—as, to that extent, it is. But in paleoanthropology, no less than in other areas of human experience, no good idea avoids the fate of being taken to a ludicrous extreme. So lumping's converse, the creation of new species, almost inevitably came to be viewed by paleoanthropologists as a bad thing in principle. And this, of course, is why *Homo habilis* had had such a rough ride to begin with. We'll look again at this matter of species; for the moment let's just note that if new species are undesirable, new genera are unthinkable!

But perhaps the unthinkable is just what we need to contemplate. By current reckoning, our family (or tribe, or whatever) contains only two genera: *Australopithecus* and *Homo*. Interestingly, at the level of the body parts that fossilize, genera seem to be the *Gestalt* category of mammalian classification: Tattersall's law states that if you can tell two skulls apart at fifty paces, you have two genera, while if you have to scrutinize them close up to tell the difference, all you have is two species. Of course, this is an oversimplification—even a grotesque one—but it does express what seems to be a basic consistency: among mammals in general, the genus is the level at which the "family resemblance" among related species expresses itself (or is perceived by us). Genera, of course, are simply collections of species all of which are descended from the same exclusive ancestor (they are thus what is called "monophyletic"); but there's obviously a limit as to how inclusive a genus can be, or every living species would belong to the same one. And if a genus is thus no more than a monophyletic group of species, why be afraid of recognizing a sufficient number of them to express the morphological diversity that has accumulated within the group? There's absolutely no reason in principle at all to shrink from this exercise; but in paleoanthropology there is one very powerful practical one: we don't have an agreed-upon cladogram in which to discern the clusters of species that could usefully be separated off as genera without violating the criterion of monophyly. With any luck, one day we shall; but for the moment uncertainty and the inertia of tradition combine to make the multiplication of genera within Hominidae a taboo subject. Too bad—but one should at least note that having to force *habilis* (and its relatives) into one or another of a mere two genera does little to enhance our understanding of the complexity of human evolution. It also makes it virtually impossible to arrive at a morphological definition of the genus *Homo* that has any substance whatever.

The most detailed analysis yet done of the *Homo habilis* situation has come from the pen of Bernard Wood. To him was entrusted the detailed description of all the east Turkana hominid crania, teeth, and jaws; and Wood, a longstanding member of the Koobi Fora team, cleaved to approved procedures in the monograph that resulted. He clearly saw that there were more species among the hominid fossils than the party line suggested, but in this massive tome he refrained from giving them names. Subsequently, however, he has been less inhibited. In evaluating the gracile fossils from Olduvai, he concluded that all could comfortably fit within the species *Homo habilis*, as defined by the fossils from low in Bed I. And if all of the Olduvai graciles belonged to *Homo habilis*, then so, almost inevitably, must OH 62—despite its extraordinarily archaic aspect. When he turned to Koobi Fora, however, Wood found a more complex picture. Some Koobi Fora hominid specimens, he concluded, were plausibly members of the same species: among these were the crania 1805 and 1813. Others however, including the famous 1470 cranium, clearly had different affinities (within the genus *Homo*) and needed a new species designation. To the chagrin of many, such a designation already existed: *Homo rudolfensis*, based on a new species of *Pithecanthropus* created by the Russian anthropologist V. P. Alexeev to contain the 1470 cranium. Matching the cranial specimens with the few postcranial bones known from East Turkana is tricky, but a couple of femora, for example, are larger and more modern-looking than those of OH 62. These might plausibly be attributed to *Homo rudolfensis*, whereas a partial skeleton from Koobi Fora called ER 3735 is more archaic, with long arms reminiscent of OH 62. Thus the few postcranial bones known seem to agree pretty well with the cranial and dental evidence that at about 1.9 myr ago two gracile hominid species existed at Koobi Fora.

Although arrived at by a loyal and expert team member, this account of the Koobi Fora and Olduvai "early *Homo*" fossils differs greatly from Richard Leakey's own preferred interpretation. Leakey's own popular account of OH 62 stressed its incompleteness (while Johanson's naturally enough emphasized the thought and diligence which was put into extracting the maximum of information from fragmentary material), yet he nonetheless seized on this specimen as proof that about 2 myr ago at Olduvai there were two kinds of nonrobust early human: the big-brained *Homo habilis* represented by the type material, and a smaller, more archaic form represented by OH 62. At Koobi Fora, these same hominids were exemplified, respectively, by the large-brained ER 1470, with which were associated the more modern postcranial bones, and by the smaller-brained (ER 1813) and more primitively proportioned (ER 3735) form. In this way, Leakey was able to recognize three distinct kinds of hominid (including robust *Australopithecus*) that were ubiquitous in East Africa in the period that preceded his "early *erectus*." And one of them was, of course, an ancient form of *Homo* that was shorn of the inconveniently primitive body proportions of OH 62.

It's too early to know what kind of pattern will ultimately be extracted from the mass of fossil material that has been attributed to *Homo habilis* at one time or another. But I think it's fairly easy to see where the trend lies in respect to the African/Asian *Homo erectus* question. Wood's study quite convincingly confirms that the "early *Homo erectus*" from Turkana represents a species distinct from that of Asia. In that case, it needs a name of its own; and the first name available for it under the rules of nomenclature is *Homo ergaster*, the name awarded by Colin Groves and Vratja Mazak in 1975 to the Koobi Fora lower jaw ER 992. As far as is known at present, there's nothing about *Homo ergaster* that would prevent it from being at least broadly ancestral to all later species of *Homo*, including *H. erectus* and *H. sapiens*; in contrast, it seems fairly certain that *H. erectus* of Java and China is not our ancestor. Still, not everyone will agree even about this, and we'll look at the controversy surrounding the origin of *Homo sapiens* in the next chapter.

Meanwhile, West Turkana had yet another surprise in store. The excavation of the WT 15000 site was still going on when, in August 1985, Alan Walker picked up part of the skull of a robust australopithecine at a locality next to another dry stream bed, this one called the Lomekwi River. This fossil, however, wasn't just another robust australopithecine; for it was 2.5 myr old, a good half million years older than any other robust known from

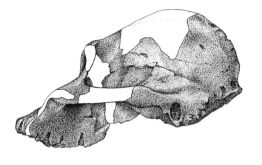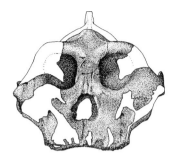

Side and front views of the "Black Skull" (KNM-WT 17000) from Lomekwi, West Turkana, Kenya. Scales are 1 cm. DM.

Kenya, and anything up to a million years earlier than *Australopithecus robustus* from South Africa. And, just as important, once Walker had pieced a more or less complete (though toothless) cranium together from numerous fragments, this specimen (affectionately known as the "Black Skull" from its dark patina, though it is more formally named WT 17000) was found to look very different from every other robust australopithecine skull known. One of the features for which all robust australopithecines then known were famous was the flatness of their faces, their shortness from front to back. Yet the Black Skull boasted a splendidly protruding, rather "dished" face with a relatively sunken nasal region. Along with the elongated face, for reasons of muscular advantage, came an equally long braincase, with a longitudinal crest concentrated toward the back. Plenty of anatomical detail, including chewing tooth roots of very generous proportions, indicated that this was indeed a relative of the younger "robust" australopithecine species; but the gestalt of WT 17000 was hardly typical for the group.

Nonetheless, when Walker and Leakey announced this find in print, they tentatively placed it as a primitive member of the species *Australopithecus boisei*. In doing this they were clearly powerfully influenced by the prevailing paleoanthropological aversion to the naming of new species; and they were equally clearly aware of that fact, since they covered themselves by noting that future finds might necessitate putting WT 17000 in a separate species. A conference on the "robust" australopithecines organized in 1987 by Stony Brook's Fred Grine provided a useful platform for discussion of this issue. At the conference the consensus emerged that WT 17000 did indeed require recognition as a species separate from *A. boisei*. But the consensus was also that a name for this species already existed—a name that Walker and Leakey had, indeed, already mentioned in their initial paper. During the international expeditions to the Omo, back in the late 1960s, the French contingent had discovered a toothless lower jaw in sediments dating back to about 2.6 myr ago. This was much older than any other hominids then known, and scrappy as the piece was, Camille Arambourg and Yves Coppens decided that it could not belong to any species (or genus—the French have always been less reticent in these matters than anglophones) that had already been described. So they called it *Paraustralopithecus aethiopicus*. Given the poor condition of the specimen this new name was widely ignored at the time; but now that a more complete specimen of about the same age had been discovered, it excited new interest. The Ethiopian jaw was from a very much smaller individual than the Black Skull and was in any case not directly comparable. But once a number of other Ethiopian specimens had been brought into the picture, as well as another lower jaw subsequently found at West Turkana, most of those attending the conference were prepared to accept a distinctive third species of robust australopithecine, exemplified by the Black Skull, which seemed to be represented by fossils ranging in age from about 2.8 to 2.2 myr old. By common consent, this form took the species name *aethiopicus*.

Another result of the Stony Brook conference was a broadening realization of just how different the "robust" australopithecines are from the "gracile" ones (quotation marks are used here because, as Grine is fond of pointing out, we don't have enough evidence of the body skeletons of the "robust" forms to know how strongly built they were overall; all we know is that they had very large chewing teeth and supporting bony architecture). Despite this, although not everyone at the conference went along, a consensus was beginning to emerge that, way back in 1938, Robert Broom had been right to place his robust specimens from Kromdraai in a genus, *Paranthropus*, separate from *Australopithecus*. Accepting this distinction, there were now three species generally recognized within *Paranthropus*: *P. robustus* from South Africa, and *P. aethiopicus* and *P. boisei* from East Africa. However, this still left the relationships among the species of *Paranthropus*, and between this genus and *Australopithecus*, somewhat up in the air. Most paleoanthropologists by then accepted *A. afarensis* as the stem species from or close to which the later hominid species had emerged, but after that opinion diverged.

Given that a robust species was now known that was as old as *A. africanus*, it was hardly plausible (though still not impossible; the full time range of an extinct species can never be known with certainty) to place the latter at the root of the robust clade, as Johanson and White had done. And it was a little iffy to make *africanus* the progenitor of all later hominids, as had also been mooted. But sheerly on grounds of time it was still possible, for example, to argue that *A. afarensis* had given rise to the East African robusts on the one hand and, via *A. africanus*, to the diverging South African robust and *Homo* lineages on the other; or to suggest that *afarensis* had simply thrown off two diverging *africanus/Homo* and *Paranthropus* lineages. The major question in early hominid phylogeny thus boiled down to whether the robust clade was monophyletic or not: whether or not the included species comprised all the known descendants of a single common ancestor. This is a problem that (despite the growing use of the name *Paranthropus*, which suggests monophyly) has still not been resolved to the satisfaction of all; and it is, of course, bedeviled by the fact that we almost certainly have our basic units of analysis wrong. If there are two things we can be sure of it's that, first, there are more early hominid species out there than we have yet, for whatever reason, been able to recognize; and, second, that among the species that we have accurately characterized, there's not one whose full time span on Earth we know.

One of the papers at the Stony Brook conference was delivered by the Yale (formerly Transvaal Museum) paleontologist Elisabeth Vrba, who had closely studied the evolution of mammalian (particularly antelope) faunas in Africa over the last several million years. She had noted that in various parts of Africa a remarkable shift in the fauna had taken place at around 2.5 myr ago: forest antelopes had become rare, to be replaced by species that graze on dry, open savannas. The conclusion was clear: some kind of climate shift had occurred that had turned much at least of the African forest to savanna. This turned out to fit neatly with geological findings indicating that a polar glaciation had occurred at just that time, decreasing average global temperatures by ten degrees Fahrenheit or more. Not only did temperatures fall, but the continents became more arid; this explained the nature of the vegetational change, which in turn accounted for the faunal shift. Vrba noted that it was about at this time, 2.5 myr ago, that *Paranthropus* appeared, that stone tools showed up in the geological record for the first time, and that the first fossil intimations of the genus *Homo* began to be seen (in the form of a skull fragment found near Kenya's Lake Baringo by her Yale colleague Andrew Hill). Were these events related? Vrba thought so. And she also concluded that they were simply part of a larger pattern that had repeated itself over and again during hominid history. For example, it was at about 5 myr ago that today's familiar antelope species began to proliferate on the African landscape, an event that also coincided with an episode of dramatic climatic cooling and drying, with a worldwide contraction of forests and expansion of savannas. And, of course, it quite possibly coincided with the emergence of the first hominoid bipeds out onto the savanna. For Vrba, then, climate shifts have dramatically affected the evolution of the human family, but as part of a much larger picture: she sees periodic changes of this kind as the cause of general "pulses" of speciations and extinctions in which our own ancestors, as well as those of a vast variety of other organisms, were inextricably caught up.

This is an attractive idea, which will live or die as we come to know more about the timing of all these varied and putatively related events. It's also one that is firmly founded in evolutionary theory, for although it undermines traditional notions of gradual change, it fits well with what we know about how speciation occurs. The fragmentation of formerly continuous populations caused by the spread or contraction of savannas (or any other habitat) produces ideal conditions for speciation. And, along with innovation at the local level, it's speciation that is the engine of evolutionary change. Though the very notion is anathema to traditionalists, "retrogressively" invoking as it does a multiplicity of extinct hominid species, it is in fact highly probable that an increased speciation rate among hominids, due to the frequent climatic fluctuations of the Ice Ages (roughly the last 2 myr) was at least partly responsible for the accelerated pace of change in the human lineage during the latter half of its existence.

15

The Cave-Man Vanishes

The remarkable finds made in eastern Africa beginning in 1959 have tended to draw attention away from South Africa, where all the excitement in the "earliest ancestor" department initially started. Perhaps in part because of their humble origin as jumbled cave fill, as well as because of the difficulties of dating that this origin presented, the South African australopithecines have somehow seemed wanting in the romance that surrounds their brethren further north. Weeks and months spent in laborious rock-bashing at a single locality like Sterkfontein tends to lack the glamour that accompanies striking out into a wild and remote landscape such as that of Hadar, where another reliably datable Lucy might lurk in every gully. Yet, without much fanfare, the inscrutable South African caves began in the 1960s to yield not only a remarkable crop of new fossils, but a story of their origins that was dramatically at odds with earlier interpretations.

A leader in this renaissance of australopithecine studies in South Africa was C. K. Brain of the Transvaal Museum. Brain was particularly interested in the way in which the cave sites had formed, and in how the bones had accumulated as part of the breccia infill. In 1965 he reopened excavations at Swartkrans, the site from which robust australopithecines were best known, and by 1970 he was able to provide a new reconstruction of the way in which the cave had originated and subsequently filled with rubble. The original underground cavity, formed by solution of the dolomite rock by groundwater, had become connected to the surface by a vertical shaft that may have descended beneath a rock overhang. Down this shaft, periodic rainstorms had washed all sorts of detritus, including dust, gravel, pebbles, and the bones of dead creatures. And whereas Raymond Dart had concluded that the broken bones found at Makapansgat had resulted from the activities of bloodthirsty australopithecines, the Swartkrans bones looked to Brain much more like leftovers from the meals of carnivores. But if so, how had so many of them found their way into an underground cave, recently reexposed at the surface by erosion?

The dolomite limestones of the generally dry Transvaal continue to be weathered today, and situations like those at ancient Swartkrans are still quite common. The entrances to vertical fissures in the rock still form depressions into which water drains, with the result that in an otherwise rather treeless landscape it is not unusual to find trees growing in such places. Leopards often take their prey up into trees in regions where the carcasses are at risk from marauding hyenas; and it was leopards, Brain proposed in 1970, that were largely responsible for the bone accumulations at Swartkrans. The australopithecines, rather than being the hunters that Dart had envisioned, suddenly found themselves among the hunted. The rare occurrence of australopithecine postcranial bones at Swartkrans by comparison with antelope bones or hominid skull fragments was attributed to the fact that primate postcranials are rather delicate; a leopard will typically consume an entire baboon, leaving only the skull, while much more will be left over from an antelope meal. In support of his idea, Brain was particularly pleased to find one partial juvenile *Paranthropus* braincase from Swartkrans that was penetrated by two identical and closely spaced holes—holes into which the canine teeth of a modern leopard jaw fitted perfectly! When dragging its prey, a leopard will often grip the victim's head in its jaws, and the *Paranthropus* braincase eloquently recounted the story of a similar sad fate.

John Robinson, as we've seen, had noticed that a few of the hominid remains from Swartkrans were much more gracile than the robust ones for which the site was most famous; such fossils included the jaw for which he had initially coined the new name *Telanthropus*, but which he had later attributed to *Homo erectus*. Now, Brain confirmed that stone tools were also present at the site. These tools were described by Mary Leakey as most closely resembling "developed Oldowan" implements from Bed II at Olduvai, though they tended to be bigger; among them were various kinds of choppers and a couple of bifaces. Unlike Olduvai, however, Swartkrans was emphatically not an activity site of the toolmakers; Brain gave cogent reasons for believing that the tools had simply been washed down into the cave along with other surface debris. Such debris had formed a conical deposit below the shaft leading from the surface down to the cavity; rocks and bones falling onto this cone were carried by their momentum to areas that might be quite distant from the opening. In this way, a very complex stratigraphy formed in the cave fill, one that took many more years to work

Artist's reconstruction of a leopard dragging away the young Swartkrans Paranthropus *whose braincase was pierced by its canine teeth. Drawing by DS after a concept by Douglas Goode.*

out, especially since it became evident that the cave had in fact been subject to several episodes of deposition and erosion.

In 1976 Brain was able to present an interpretation of the geology of Swartkrans by the Chicago geologist Karl Butzer that divided the breccia fill into two members (lithologically distinct units): Member 1 (with lots of hominids) and Member 2 (with many fewer hominids, but most if not all of the stone tools). Later excavations revealed a more complicated situation, however, and Brain ultimately divided Member 2 into four, with hominid fossils known only from the lower part of that sequence (Members 2 and 3). Dating of these various geological units (and of those at other South African cave sites) was also given a boost during the 1970s and 1980s by comparisons of their contained faunas with the newly available dated faunal sequences of eastern Africa: Basil Cooke was a pioneer in this, and studies of the antelopes by Elisabeth Vrba and of the monkeys by Eric Delson of the City University of New York and the American Museum of Natural History proved particularly useful. Such faunal comparisons suggest that Swartkrans Members 1 and 2 fall into the time range of about 1.9 to 1.6 myr ago, with Member 3 somewhat younger, around 1.5 to 1.0 myr old. Both *Paranthropus* and *Homo* fossils are known from Members 1 and 2, while only *Paranthropus* is known from Member 3 (though sparsely). It's notable, however, that while Swartkrans has produced the fossil remains of dozens of *Paranthropus* individuals, only six *Homo* fossils are known. Stone tools occur throughout, as do pieces of antelope bone (mostly long bones and horn cores) that have acquired a polish typical of bone artifacts

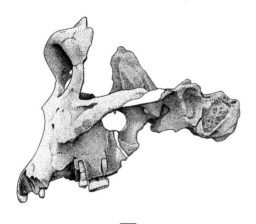
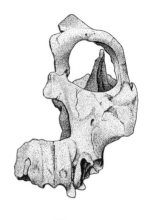

Side and front views of the reconstructed partial cranium SK 847, from Member 2, Swartkrans, South Africa. Scales are 1 cm.   DM.

that have been used for digging in the ground for roots, tubers, and so forth. The amount of wear on many of these is characteristic of prolonged use, and thus suggests that they were carried around for days at a time.

Who made or used these stone and bone artifacts? Either hominid might have; but most paleoanthropologists feel that while it may well have been *Paranthropus* that used the bones for digging, it was *Homo* that made the stone tools. Randy Susman, however, who has studied hand bones from Swartkrans that were found during Brain's excavations, has concluded not only that they indicate a capacity for precision gripping equal (if not superior) to that evinced by *Homo habilis* at Olduvai, but that they belonged to *Paranthropus*. If this is so, *Paranthropus*, by far the most abundant hominid at Swartkrans, might plausibly be the stone tool maker. Susman's interpretation has caused considerable controversy, however, and for the moment the authorship of the stone tools from Swartkrans remains in question. Interestingly, though, despite the fact that Member 3 has produced only australopithecines, it is from here that traces of fire are known. These occur in the form of burned stones and bones, heated to temperatures typical of campfires. Member 3 times were the only point in the early history of the cave at which hominid occupation of the cave entrance might have been possible, and this might account for the fact that burned objects occur only in that member; however, Brain prefers the idea that the introduction of fire took place between Member 2 and Member 3 times. As to the fire user (for Brain is as reluctant to conclude that a fire *maker* was involved as he is to affirm that the fire was used in cooking), few doubt that, despite the lack of fossils, it was the gracile hominid ascribed to *Homo*—but to what species of *Homo*?

In 1970 Ron Clarke had realized that a Member 1 palate, ascribed to *Telanthropus*, actually joined with a facial fragment and a portion of temporal bone (both of which had been assigned to *Paranthropus*) to form part of the left side of the cranium of a single individual of *Homo*, dubbed SK 847. The species assignment remained in doubt, however, until Alan Walker visited South Africa in 1977, bearing a cast of the newly discovered KNM-ER 3733. The close resemblance of the two specimens struck both scientists, who agreed that they thus had before them an example of *Homo erectus*. This, of course, brought them into line with John Robinson, who had settled years earlier on a similar attribution for his *Telanthropus*. However, the question was opened again when doubts began to be voiced about whether the "early *erectus*" of Africa should actually be placed in that species; and a recent reconstruction of SK 847 by Fred Grine, employing a sophisticated computer imaging technique, has highlighted some significant differences from *Homo ergaster* of East Africa. For the moment, then, the species identification of Swartkrans *Homo* must remain in doubt: but since it doesn't closely resemble anything allocated to *Homo habilis*, it will probably end up being placed in a new species closely related to *Homo ergaster*.

In parallel with the work at Swartkrans, activity was also picking up at Sterkfontein. In 1966, during the celebrations for the centenary of Robert Broom's birth, Phillip Tobias and Alun Hughes presented detailed plans for reopening excavations at this classic *Australopithecus africanus* site. This work, which continues today, had resulted in the collection of hundreds more hominid fossils (most, alas, fragmentary) by the time of Hughes' death in 1991. Most of these fossils are as yet undescribed, for Tobias dedicated the quarter-century following the discovery of Zinjanthropus to the massive labor of monographing the Olduvai fossils, leaving until later the equally monumental task of documenting the Sterkfontein collection. Notable exceptions, though, include the Stw 53 cranium discovered in 1976 and attributed to *Homo habilis*, and a few other fossils that were described by Ron Clarke during the 1980s. These included a heavily reconstructed cranium (Stw 253), which Clarke interpreted as showing a number of features uncharacteristic of *A. africanus*. Unlike Stw 53, which came from the site's later Member 5 (which its fauna suggests is about 1.6 myr old, and which also has yielded stone tools) this new skull came from the earlier Member 4 (about 2.5 myr old), the deposits that had yielded the classical suite of *A. africanus* fossils. Reexamination of other Member 4 specimens convinced Clarke that there were, indeed, two kinds of hominid present in these early sediments: a smaller-toothed form (classic *A. africanus*), with a rounded brow and a prominent nasal region, and a larger-toothed one (including Stw 253) with a flatter brow and midface. Rejecting for solid reasons the notion that these two forms might merely be males and females of the same thing, Clarke concluded that two lineages were represented at Sterkfontein. Despite its rather expanded front teeth, Clarke discerned in the larger-molared form a precursor of *A. robustus*. As one sees more of this material it becomes plainer that Clarke is right in noting more than one species in Member 4 of Sterkfontein; but it seems likely that as more fossils are described and analyzed the picture will become more complex than even he has yet articulated.

Despite the relatively small number of hominid fossils known from Makapansgat a similar pattern seems to obtain there, with both smaller- and larger-molared individuals represented. Once again, reanalysis of the fossil bone assemblages suggests that they are the work of carnivores and scavengers, rather than of australopithecines wielding osteodontokeratic tools. With one single exception, all of the fossil hominid remains at Makapansgat come from the "Grey Breccia" deposits known as Member 3; the associated fauna suggests a date of around 3.0 myr ago, making Makapansgat the oldest of the South African australopithecine sites. The age of the original Taung skull remains a mystery; monkey fossils found at about the same time suggest an age of about 2.2 myr, but there is no guarantee that all came from the same deposits. If this date is correct, however, the Taung baby is geologically as well as developmentally the youngest of all of the South African fossils attributed to *A. africanus*.

The South African australopithecine sites thus seem to cover a span of as much as 1.5 to 2.0 myr, in the following sequence (from oldest to youngest): Makapansgat 3, Sterkfontein 4, (?Kromdraai, ?Taung), Swartkrans 1, Swartkrans 2, Sterkfontein 5, Swartkrans 3. Specimens allocated to *A. africanus* come earliest in the sequence; *Paranthropus* appears next, at around the same time as *Homo*. During the long period covered by these sites, southern Africa appears to have been undergoing a general drying trend; in Makapansgat and Sterkfontein 3 times the vegetation of the high veld was probably rather bushy, with forest lining streambanks, while by Swartkrans times open savanna seems to have prevailed. Could this climatic shift explain the different dental adaptations of the robust and gracile types? Early on, John Robinson had suggested that the differences between the teeth of *Australopithecus* and *Paranthropus* were to be explained by diet: the gracile dentition was adapted for omnivory, while the robust one was that of a highly committed herbivore. More recently Fred Grine, along with Rich Kay of Duke University, has used electron microscopy to study the wear produced by chewing on the occlusal surfaces of the molars of the two types. Kay and Grine found that in *A. africanus* from Sterkfontein the tooth enamel was polished and lightly scratched, in contrast to the heavy pitting and gouging seen in *Paranthropus* from Swartkrans. Clearly, the two forms were eating different things, the latter specializing on harder, grittier substances: exactly the kind of vegetable edibles, such as roots and tubers, that are found out on the open savanna—and that digging implements are particularly helpful in obtaining. However, *A. africanus* also turned out to be a vegetarian, simply feeding on different kinds of plant products, perhaps mostly fleshy fruits.

Whether this correlation between climate, vegetation, and morphology is cause or effect remains to be clarified. Meanwhile, C. K. Brain believes that the gap between Members 4 and 5 at Sterkfontein (approximately the million years following 2.5 myr ago) represents a critical period of human evolution, one in which the hunted indeed became the hunters. His investigations suggested to him that during the time when Member 4 was being laid down, the cave entrance was a lair of carnivores, which dragged their australopithecine victims into its recesses. By Member 5 times, however, "the men had not only evicted their predators, but had taken up residence in the very chamber where their ancestors had been eaten." Yet, Brain added, "they were not more than amateurs in hunting . . . the nature of their antelope remains . . . suggests that they depended heavily on the kills of professional carnivores *before* they progressively developed their own prowess as hunters." The tricky nature of the South African cave sites makes it difficult at this stage of the game to go much further in assessing the capabilities of the earliest *Homo*. In East Africa, however, conditions for investigating this problematic issue are much more favorable—although consensus among archaeologists is nonetheless elusive.

As work progressed at such East African sites as Olduvai and Koobi Fora, the range of questions that the archaeologists asked of the material at their disposal considerably broadened. Among other things, attention began to shift from the shape of the artifacts found to the technology that produced them. For the experimental production of stone tools by archaeologists taught them that the shape of a stone implement results at least as much from the form and nature of the piece of rock chosen to start with as from the intentions of the toolmaker. So if you want to know about what was going on in the toolmaker's mind (and archaeological evidence for behavior tells you a great deal more about that than does the size and outside shape of his brain), the manufacturing process is much more informative than the end product. Mary Leakey, for example, put a great deal of effort into sorting Oldowan tools into a large number of different categories, identifying a whole "kit" of implements: spheroids, polyhedrons, discoids, choppers, and so forth. Most of these consisted of one or another kind of modified "core": cobbles off which flakes had been struck. The assumption was that the cores—the pieces of stone that had actually been modified—

were the implements that the toolmakers had intended to produce. Experiments, however, showed that quite likely it was the sharp flakes knocked off in this process that were the actual tools used for cutting. The idea that the different core types represented "mental templates" in the heads of the toolmakers did not seem to be borne out: they were simply by-products of varying amounts of flake production using cores of different shapes, sizes, and materials.

But did this mean that that the Oldowan hominids, the makers of the earliest stone tools, were simply opportunists who struck flakes from whatever pebbles happened to be handy where cutting tools were needed? The answer to this one appears to be no. Mary Leakey had noticed early on that Oldowans had carried suitable rocks quite a distance to the places where she found the tools made from them. As it turned out, these early toolmakers were not highly selective; although they collected rocks that were suitable for toolmaking, they didn't everywhere make a great effort to amass the best possible materials. But at many localities in the Koobi Fora region, for example, the nearest natural sources for the lava cobbles turned into tools at archaeological sites turned out to be several kilometers away. Hominids must have carried these raw materials in over such considerable distances, and that flaking took place on site is shown by the fact that often flakes found close together can be joined up to reconstruct an intact core. Moreover, it is not uncommon for archaeologists to find more than one type of "foreign" rock at a given site, indicating that such objects had been brought in from several distant points on the landscape. Such activities on the part of early hominids require a degree of forethought quite out of the range of living apes, who, on the rare occasions when they make tools—most famously the stripped twigs used in "fishing" for termites—pick up the raw materials at the spot where they are used.

Experiments carried out by Nick Toth of Indiana University suggest the same thing from another vantage point. Lumps of rock battered into roughly spherical shapes are quite commonly found worldwide at Stone Age sites, sometimes in large numbers. What these apparently deliberately shaped objects were made for remained a puzzle for many years, although they were often seen in earlier times as "bolas stones," tied together by thongs and thrown to entangle the legs of prey animals. By experiment, Toth was able to show that almost any lump of stone he chose would assume this spherical form after many hours of being banged against other pieces of rock. The distinctive shape was thus a passive result of use as a hammer, reflecting no intent on the part of the toolmaker to produce a spheroid. Once more, then, the "mental template" idea failed. On the other hand, this finding did nothing to diminish the fact of the toolmaker's intent. For it took far more than a single session of tool making to produce a spheroid, implying that the toolmakers habitually carried their favored hammerstones around with them from place to place in anticipation of needing them.

One interesting experiment performed by Toth, his Indiana University colleague Kathy Schick, and a group of psychologists at the Yerkes Primate Research Center concerned the ability of a living ape to make and use tools. Noting that it had become fashionable to look upon the early bipeds— roughly, anything prior to *Homo ergaster*—as "bipedal apes", Toth, Schick, and colleagues tried to determine how far a bonobo ("pygmy chimpanzee") could be trained to flake simple stone tools. Their subject, Kanzi, a star in communication experiments, showed an immediate interest in having sharp flakes available to cut cords that held a fruit-containing box closed. He got the idea of striking flakes from a core, but even after many months of training he was still nowhere near the skill level of the Oldowan toolmakers. The latter clearly understood the major properties of the stones they worked and selected the most effective points at which to strike an inevitably irregular core. Not so Kanzi, who never mastered the idea of striking stone at the optimum angle. His best products are rather like the "eoliths" that so confused early archaeologists: rocks randomly banged together and flaked as they rolled along riverbeds. Toth and colleagues concluded from this that the early hominid toolmakers had a much better cognitive understanding of what toolmaking is all about than any modern ape is able to acquire. And from that they hazarded that in hominid prehistory there must have been a stage of stoneworking that preceded the Oldowan, but that by its nature it would be difficult or impossible for archaeologists to identify or to discriminate from the results of natural forces. Still, these experiments give some idea of what one might need to look for.

Another aspect of experimental archaeology centers around how archaeological sites are formed, and how they may later be disturbed and thus distort the story they contain. An archaeological site is just a place where there is evidence of early human activity, and the nature of the site depends on what activities were carried out there. In the early Paleolithic the range of hominid activities preserved tended to be pretty limited, boiling down essentially to the making of crude stone tools, or to the butchering of animal carcasses, or both. Each of these processes can to some extent be mimicked by natural forces, particularly by water action which tends to concentrate objects that are simply lying around on the ground. The creation of "artificial" archaeological sites by stoneworking and butchering at various points on the landscape can help in recognizing anthropogenic concentrations of stone and bone, whether pristine or altered by natural forces. Nonetheless, there is considerable disagreement over what most Oldowan sites tell us about how the earliest toolmakers were making a living.

Most Oldowan sites contain the bones of a variety of mammal species and body parts, some bearing cut marks made by stone tools, or fractures that might well have been caused by hammering at them with a stone to get at the marrow. In earlier times this was taken as evidence of significant hunting prowess: the ability to kill animals of sometimes considerable size. But Rick Potts and Pat Shipman noted in 1981 that on bones from Mary Leakey's localities in the lower levels at Olduvai cut marks were often made over grooves already left on the bones by the teeth of carnivores, indicating that the carnivores had got to the carcasses first. What's more, Lewis Binford realized at around the same time that these sites also contained a preponderance of bones from body parts that bore little meat. From this and other evidence Binford concluded that the putative killer *Homo habilis* had in fact been a scavenger, attacking what was left of carcasses that carnivores had killed, fed on the choice parts of, and abandoned. On the other hand, analysis of the cut marks left on some of the bones suggested to the archaeologists Henry Bunn and Ellen Kroll that stone tools had in fact been used to dismember the higher-yielding parts of the animals. If this were indeed so, the hominids who did the job either must have freshly killed the animals they dismembered or were effective enough to chase away the carnivores who had. Bunn and Kroll's reading of the evidence has, however, been disputed on a number of grounds, and if pressed, most archaeologists would probably plump at present for a relatively humble scavenging role for the first stone tool makers. Or they would at least remain agnostic on the matter. To place all this in perspective, however, it is wise to bear in mind that, even in recent times, meat has tended to make up only a minor proportion of the diet of hunting and gathering humans.

What about the localities themselves? The concentration of bones and artifacts at some well-preserved early Stone Age sites certainly suggests that ancient hominids returned to them repeatedly, although their reasons for doing this are still obscure. As we've seen, Glynn Isaac's early notion that they represent places where food was brought to be shared out has been more or less abandoned. But, as Kathy Schick pointed out, it still appears that these locations were "favored" by early hominids. Perhaps such sites were simply centrally located within group territories; or maybe they offered shade or sleeping trees, or a good view of the surrounding landscape and potential predators; or maybe they fulfilled some more specifically social function. Some sites, it's been suggested, may have served as depots for stones suitable for flaking; this would have minimized the carrying of such items that would have been necessary. And at the simplest end of the spectrum, other sites may simply have been places where animals were butchered where they lay. What seems sure is that there is no good evidence anywhere for structures of any kind. The stone circle at Olduvai site

DK may well have been caused by fracturing of the underlying lava by the roots of a tree, and is in any case rather indefinite; and although similar phenomena have been reported from elsewhere, there really are no convincing candidates for hominid-built structures until very much later in the archaeological record. Fire, just possibly, is something else. A couple of sites in Kenya, one at Koobi Fora and another at a place called Chesowanja, show areas of baked clay resembling what typically forms below campfires; both sites are about 1.5 myr old—as old as or older than the burned bones and cobbles from Swartkrans. However, it remains possible that the Kenya occurrences result simply from wildfires, and there is certainly no evidence of organized cooking hearths this early in the record.

I've already noted that at about 1.5 myr ago, some time after the appearance of *Homo ergaster*, handaxes and cleavers appear in the archaeological record, ushering in the Acheulean industry. Finally, here is a tool type that was unquestionably made according to a "mental template" that existed in the minds of the makers. For handaxes were no chance result of flake production; at some sites these bifacially flaked and carefully shaped tools have been found in extraordinary abundance, with a remarkable consistency of size and shape. Although rather unwieldy when large, and sometimes very heavy, handaxes represent an astonishingly successful technology, spreading throughout all inhabited areas of the Old World apart from eastern Asia, and remaining in production for well over a million years. Exactly what kind of cognitive advance the making of such tools represents is not clear. Indeed, it is not at all obvious that cognitive advances (that is to say, improvements in conceptualizing abilities) go hand in hand with technological advances; after all, the latter have ultimately to result from an innovation by an individual, who is unlikely to differ wildly in cognitive capacity from his—or her—parents. Any technological advance, in other words, has to be within the cognitive abilities of individuals belonging to the species concerned; it can't enlarge those abilities.

Experimental work by Peter Jones and by Toth and his colleagues suggests that the handaxe form was best adapted to the task of butchering large animals (though it has been shown that simple Oldowan flakes are capable of cutting through the inch-thick skin of an elephant, and microscopic analysis of the worn surfaces of such tools has shown that they were used for cutting meat and soft plants, as well as for working wood). Picks (slender-tipped handaxes) seem to be particularly good for digging. Nonetheless, despite the utility of handaxes in butchery, there is still some question about the lifestyle of the handaxe makers. Traditionally, such handaxe sites as Spain's Torralba and Ambrona have been viewed as places where very large mammals were killed; but more recent studies have emphasized the role of carnivores and other natural forces in creating the assemblages.

Similarly, eastern African handaxe sites with the remains of large mammals can often be explained by factors other than hominid predation. The upshot of new excavations and new analyses of older evidence has been an outbreak of archaeological fence-sitting on the matter of the hunting prowess of Acheulean hominids. Particularly interesting in this connection is Binford's analysis of the bone assemblages from Zhoukoudian, which, as you'll recall, Franz Weidenreich had interpreted as the result of hominid carnivory and cannibalism. According to Binford, the activity of hyenas was particularly important in creating the bone accumulations at the site and may even have been responsible, along with geological factors, for the broken-up condition of the human remains.

Binford's interests also extended to later periods of human evolution. While studying the "Mousterian problem" he became convinced that modern hunters and gatherers were an inappropriate model for trying to understand Neanderthal lifestyles which, he believes, were in fact entirely different from anything evinced by modern humans. Analyzing bones and tools from levels dating from approximately 125 to 70 kyr ago at the site of Combe Grenal, in western France, Binford found that at each level two separate concentrations of artifacts and bones occurred. In "nest" areas were found plenty of ashy materials that indicated that fires had burned there (though there were no hearths), plus plenty of simple flake tools made from local stone and marrow bone fragments. Elsewhere were scattered smaller concentrations of bone, with more sophisticated stone tools, such as retouched scrapers, that were made from materials brought in from distant localities. What's more, the animal bones associated with the scrapers were regularly those of species that lived in environments where the stone came from; the conclusion was that the food remains also had to have been carried in from some considerable distance away. Binford hazards that the nests were where females lived, and that the scraper sites were made by males; to cut a long story short, if true, this suggests that males and females led largely separate lives, the males ranging widely and returning only occasionally to join the the females, who led more sedentary existences. For Binford, there is no evidence at Combe Grenal that Neanderthals lived in families, reproductive units in which resources were shared among all members.

Binford also infers differences from modern people in the distribution of Neanderthal sites. These do not occur in areas of extensive grasslands where vast migrating herds moved great distances but in predictable patterns, and where they were widely exploited by early modern people. Instead, Neanderthal sites were concentrated in areas of varied vegetation where the resources, if more limited, were also more constant and required less foresight in their exploitation. Systematic hunting of large-bodied mammals, Binford believes, is a monopoly of behaviorally modern *Homo sapiens*. Whether or not they accept the rest of Binford's analysis this, at least, is a conclusion which other archaeologists increasingly share.

The archaeological record shows clearly that the Neanderthals were less inventive, less innovative, than the modern humans who replaced them. But there's no denying that, like us, they had large brains. Does this imply that even if they were rather unimaginative, they possessed other human features such as language? It turns out that neither the size nor the external appearance of the brain is of much use here: there is simply no way of reading function with adequate precision from the bumps and fissures on the outside of the brain (and still less from brain casts). So no help is forthcoming from that direction. Speech, however, is a (somewhat) different matter from language as such. For, to produce the sounds that are associated with modern articulate speech, you need specialized anatomical equipment apart from the brain. Notably, you have to have a larynx (voicebox) that is situated low in the throat, connected to the oral cavity above by a long section of tubing (the pharynx). This long pharynx is manipulated by the muscles of the throat to modulate the vibrations produced at the larynx, and thereby to make the basic sounds on which articulate language depends. Primitively, the base of the hominoid (indeed, mammal) skull is flat. This reflects the presence of a high larynx and a short pharynx, limiting the range of sounds that can be made. Among modern humans in contrast, space for a high, looping pharynx is created by bending the base of the skull downward, creating a characteristic flexion.

In the early 1970s, the anatomist Ed Crelin and the linguistician Philip Lieberman had the idea of reconstructing the vocal tracts of fossil hominids using the shape of the skull base as a guide. Using Marcellin Boule's original reconstitution of the La Chapelle Neanderthal skull as a first example, Crelin made a model of the airways; Lieberman then used this in a computer simulation of the sounds that it could produce. The model proved deficient in three of the most basic sounds associated with articulate speech. This work was subsequently extended and refined, notably by Crelin's former student Jeffrey Laitman, now of New York's Mount Sinai School of Medicine, who has noted a trend among the hominids. Among the australopithecines, the skull base is flat, just as it is in apes and all other mammals. But in *Homo ergaster* there is a slight but measurable flexion; and the 150 kyr-plus Kabwe skull (if not that from Petralona) looks almost modern in this respect. The Neanderthals, however, buck the trend. A recent reconstruction of the La Chapelle skull does show more flexion than the original Boule version; but it's definitely much less than what we see in Kabwe. A similar story is told by the La Ferrassie skull, also about 50 kyr old, although certain earlier Neanderthals, such as that from Saccopastore in Italy (ca. 100 kyr), appear to have somewhat more flexed crania.

Lowering of the larynx to permit refined sound modulation is not an unalloyed blessing: the high-larynx configuration permits simultaneous breathing and swallowing, and this eliminates the possibility of choking to death, an inconvenience to which modern humans are regrettably subject. Changes of the airways in the modern direction thus involve a distinct trade-off, so we're clearly in a gray area here and more information is needed. One possibility is that the apparently primitive high larynxes of the putatively cold-adapted Neanderthals were in fact specializations: one way of dealing with consistently cold, dry inspired air. Well, maybe. Hope, it must be said, lingered until recently that the discovery of fossil human hyoid (throat) bones would help with the problem of early speech by providing more direct evidence of throat structure. However, a beautifully preserved Neanderthal hyoid belonging to the Kebara skeleton has succeeded only in declenching an appropriately vociferous argument over its significance. What there is of this bone looks pretty modern, but the problem is that only a small part of the whole hyoid actually ossifies; what this element's long-disappeared cartilaginous portion looked like is anyone's guess. However the hyoid argument works out, however, when you put the skull-base evidence together with what the archaeological record suggests about the capacities of the Neanderthals and their precursors, it's hard to avoid the conclusion that articulate language, as we recognize it today, is the sole province of fully modern humans.

16

Candelabras and Continuity

For most of the second half of the twentieth century the paleoanthropological limelight was almost completely monopolized by the search for our earliest hominid ancestors. The origin of our own species, *Homo sapiens*, though of equal or even greater intrinsic interest, just didn't seem to have the same inspirational power. Perhaps this was in great part because paleoanthropologists of the period weren't able to agree on what *Homo sapiens* is (or was), although most of them were prepared to include in this species a motley assortment of fossils going well back into the middle Pleistocene, a half-million years ago or more. And in part, I suppose, it was due to the fact that the adherents of the Synthesis simply didn't expect to find the origin of *Homo sapiens* in an event as such. Rather, they expected a slow transformation, in the long course of which it would be impossible to identify any single point at which full humanity emerged. In retrospect, it seems almost inevitable that it was to the elaboration of this notion that the proponents of the single-species hypothesis retreated, once it became no longer possible to deny that in earlier times two and probably more hominid species had coexisted in Africa. And in doing this, they performed the valuable service of drawing attention back to the question of modern human origins.

You'll recall that before World War II Franz Weidenreich had developed the theory that the various major modern groups of mankind (he explicitly recognized four: Australian, Mongolian, African, and Eurasian) had distinct origins going back to the time of *Pithecanthropus* and beyond. Each of these lineages had evolved independently, at its own pace. But if so, how had they all managed to remain members of the same species? Weidenreich had the answer. "The tendency to transmute the primitive types into those of recent man," he wrote in 1939, "must be considered inherent to the form as such." In other words, Weidenreich invoked a form of orthogenesis—an innate urge to evolve towards a particular goal—as a mechanism to explain how several distinct human lineages had managed to change independently yet arrive at more or less the same point. Even in prewar China this explanation must have sounded more than somewhat old-fashioned; but Weidenreich was simply seeking a justification for something that he believed he saw in the fossils. In terms of those fossils themselves, Weidenreich's "Australian group" was the offspring of a lineage, ultimately descended from *Gigantopithecus*, which passed from Meganthropus, through Pithecanthropus, and reached early modernity with Dubois's Wadjak cranium. *Gigantopithecus* also gave rise to a lineage that passed from Sinanthropus, through a series of unknown intermediates, to the Upper Cave skulls from Zhoukoudian, and ultimately to modern Chinese and other eastern Asians. Modern southern Africans stemmed from a series of predecessors that included the Broken Hill cranium and, later on, Broom's Boskop skull. Finally, in Europe and western Asia the sequence ran from unknown precursors through the Tabūn Neanderthals, the Skhūl early moderns, and the Cro-Magnon remains, before culminating in modern Eurasians.

For some reason, Weidenreich chose to represent all this in a virtually unreadable, densely gridded, and regularly geometric diagram that seemed to suggest that many more parallel lineages had in fact existed alongside the basic four that he named, with diagonal connections at regular intervals between them all. This diagram is at once elegant geometrically and of a mind-boggling complexity biologically; and like any formulation in which every possible combination of points is joined by some combination of lines, it is susceptible to fundamentalist interpretation in a vast variety of different ways. At a recent conference in Jerusalem I was astonished to find myself attending a session devoted to the vigorous exegesis not, as I'd expected, of whether Weidenreich had been right or wrong, but of how badly Harvard's Bill Howells had misconstrued Weidenreich in his book *Mankind in the Making*. In this volume Howells had dubbed Weidenreich's view the "candelabra" model of human evolution, based on a simplified version of Weidenreich's diagram that looked something like a candlestick bearing four candles. He contrasted this to the alternative "hat rack" theory, where a single central stem sprouted off a few side branches. In the candelabra the candles, or parallel lineages, ran just as Weidenreich had indicated; Howells's sin was to have made the diagram readable by eliminating the diagonals of Weidenreich's original. Many of those present at the conference seemed to be greatly upset by this for, even though Weidenreich himself had not been excessively concerned with genes, his later disciples were. And, as we'll see in a moment, the diagonals (meaningless in the original, as far as I can tell) had now become the key to their new interpretation of the master's pronouncements. Perhaps Jerusalem was an appropriate locale for this scientific farrago; the tone of the discussion was positively theological, and it brought home to me very clearly how ingrained received scientific wisdom can become, as well as how important it is to find a respectable pedigree for one's ideas.

Now, Howells is today a much revered elder statesman of paleoanthropology; and the vehemence of the attacks on his candelabra diagram (more than thirty years after the fact) may, I think, have been motivated much less by its substance and authorship than by the fact that it coincided with the version of Weidenreich's views propounded (and inadvertently discredited) by the University of Pennsylvania anthropologist and noted TV personality Carleton Coon. In their original form, it must be said, Weidenreich's notions had attracted rather little attention; as Coon himself said in the introduction to his *The Origin of Races* (1962), "Like other premature comets of science, Weidenreich's idea flashed across the sky and was gone, obscured by the clouds of incredulity released by his fellow scientists." These incredulous fellow scientists, Coon continued, believed that "the living races of man could have become differentiated only after the stage of *Homo sapiens* had been reached." Coon thought otherwise, and he exhaustively scrutinized virtually every human fossil then knownin an attempt to show that five distinct human "racial lines" (add the Bushmen of southern Africa to Weidenreich's four) could be traced back as far as the origins of the genus *Homo* itself—as Coon understood it over 30 years ago, back to *Homo erectus*. But although he demonstrated to his own satisfaction (and, by his account, to that of two of the giants of the Synthesis, Ernst Mayr and George Simpson) that human subspecies could be older than the human species, few other paleoanthropologists agreed. Indeed, poor Coon—who had actually compiled a document that for all its deficiencies still reads impressively today—was widely and unfairly reviled for propagating a racist doctrine. The strength of this reaction was understandable, due as it was in great part to an emotional rejection, in the liberal 1960s spirit, of the notion that *Homo sapiens* was anything but the most closely knit of species. Less viscerally, though, it also reflected the perception that different lineages could hardly evolve separately into the same new species.

The objectors were actually right on both accounts; but, as it ironically transpired after the furor over Coon's book had died down, it was precisely the inheritors of the good-hearted liberal tradition, the guardians of the Synthesis which had seen such anachronisms as orthogenesis off the stage, who were to resuscitate Weidenreich's ideas. They managed to distance themselves from the odium of Coon and the candelabra by returning to the pristine purity of the founding document, in which they reinstated the diagonals of Weidenreich's original diagram. These, they claimed, represented gene flow between adjacent populations, even as local lineages followed their own independent evolutionary paths. Genes were, it was said, exchanged between neighboring lineages in sufficient quantities to ensure that all remained part of the same big happy species.

Thus was born the "multiregional continuity" industry that has employed so many paleoanthropologists in the years since the demise of the single species hypothesis, and that stands as an enduring testament to the bewitching power of the neodarwinism of the Synthesis. The most cogent early statement of the idea of multiregional continuity was put forward by Alan Thorne of the Australian National University and Milford Wolpoff of single-species fame, in an article published in 1981. In this work Thorne and Wolpoff argued that a distinctive regional population could be traced in Australasia for almost a million years, from the time of the Javan Sangiran 17 specimen to early modern Australians from the Kow Swamp site (ca. 10-14 kyr). In later publications these authors, with their students and collaborators, have broadened the original idea of continuity in island Asia and Australia to embrace other regions, most notably China but also virtually everywhere else in the Old World. The basic idea underlying all this is that when hominids first emigrated from Africa around a million years ago, they spread throughout Eurasia. As these émigrés (*Homo erectus* by Thorne and Wolpoff's reckoning at the time) encountered unfamiliar environments in their new homes, they rapidly evolved area-specific adaptations to help them cope more successfully with these unaccustomed environments. In this way, regional physical distinctions became rapidly established; these then lingered over a vast period of time, even as gene exchange between adjacent populations helped keep all hominids united in a single species. Even the rather daunting prospect of explaining how different stocks could evolve separately into the same new species turned out to be no problem for Thorne and Wolpoff: when they finally realized the true difficulty that the species boundary represented, they found a solution in the simple expedient of following Mayr's advice and making *Homo erectus* simply a primitive form of *Homo sapiens*.

If the tone of this account seems slightly less temperate than most of what has preceded it, please forgive me. I hope I have managed not to mangle the multiregional continuity idea any more than extreme compression necessarily requires, and I hope even more that I have not given the impression that its exponents are anything other than highly capable and knowledgeable scientists. But this viewpoint does seem to me to illustrate, better than any other current example, the extreme parochiality with which paleoanthropology is cursed. Up to this point, it appears, we paleoanthropologists have proven unable as a group to shed our major historical burden: the birth of our science out of the study of human anatomy, rather than out of the comparative anatomy and geology from which other areas of vertebrate paleontology emerged. Under the dead hand of this heritage, which places our species at the center of the academic universe, we seem incapable of seeing *Homo sapiens* as simply one more species among many. We constantly seek special explanations for ourselves. And, perhaps worst of all, we are afflicted with a tradition of looking for variability in the collections of fossils, rather than for diversity. This may sound like a minor quibble, but our "search image" is actually crucial to the way in which we interpret the fossil evidence of our past. Variability (within species) and diversity (among them) are not simply two sides of the same coin. As human anatomists, we are acutely aware of the fact that *Homo sapiens* is highly variable, both within and among populations, in virtually every physical trait. And we enthusiastically overextend this yardstick into the business of classifying our fossils. Add the conventional gradualist view of human evolution to an exquisite sensitivity to anatomical variation, and it's not surprising that Thorne and Wolpoff didn't have a problem in squeezing *Homo erectus* and *Homo sapiens* into the same species; after all, the revered Ernst Mayr hadn't, and to the traditional mindset the boundary between ancestral and descendant species is anyway purely arbitrary.

Paleontologists in other subdisciplines have a different perspective, however. For in the nature of things they are concerned with lots of species. Nobody, except perhaps an exterminator, can make a career studying one species—or one genus, or one family—of rodent. Paleontologists studying nonhominids must quite obviously be concerned with diversity: the extraordinary variety of species which evolution routinely throws up within every successful major group. Diversity is an inescapable fact of nature—unless you happen to be obsessed with one species alone. Yet just because there is only one hominid species in the world today, are we justified in concluding that there has only ever been one? The fossil record at Koobi Fora tells us otherwise, and I shall argue later that it is only with the arrival of ourselves—behaviorally modern *Homo sapiens*—that an entity of a truly unusual kind appeared on Earth. For now, it's sufficient to note that there's no good reason to look at our fossil precursors with eyes any different from those we focus on any other mammal species. And it was in the spirit of that realization that, in the mid-1980s, I began to look again at the human fossil record.

One thing that was bothering me at the time was the problem of recognizing species in that record. Species are the basic unit of evolutionary analysis, making the grouping of fossils into species the most fundamental process in paleontology. If we get our species units wrong, our further analyses will be invalidated all the way down the line. But there is a basic quandary here. For speciation—the establishment of definitive genetic isolation between related populations—is an event that does not necessarily have anything to do with morphological change, certainly at the level at which we are able to detect it in the fossil record. Little as we know about the actual mechanisms of speciation, it's clear that they involve genetic events which have to do with reproductive compatibility at some level or another, and not with adaptation as such. In other words, whatever its underlying mechanisms, speciation is not simply a passive result of morphological change under the guiding hand of natural selection. This being so, a species may on the one hand accumulate a large amount of adaptive or random morphological variation while still retaining its reproductive cohesiveness. On the other, however, that cohesiveness can quite easily be disrupted in the absence of appreciable morphological diversification. Pity the poor paleontologist; for if speciation has nothing necessarily to do with bony or dental morphology, which is essentially all that the fossil record offers to help in species recognition, what is he or she to do?

The only reasonable yardstick is to look not so much at the variation that accumulates *within* species (for all species will be variable, and in closely related species ranges of variation will be likely to overlap pretty much totally in most characteristics), but at the kind of variation you typically find *between* closely related species. This latter, as we've seen, is precisely the kind of thing paleoanthropologists have not been accustomed to doing much of. And when they do turn their attention beyond *Homo sapiens* they traditionally focus on our closest relatives, the apes. After all, as our nearest kin, the apes surely have most to tell us about ourselves. There are two problems with this, however. One is that the great apes, while undeniably our closest *living* relatives, are not actually all that close, for while most other living primate species have relatives within their own genus, the apes lie well outside ours. More significantly, though, today's great apes, members of a group that's been steadily declining since the later Miocene, are not very diverse at all: there are only one species each of orangutan and gorilla and a mere two species of chimpanzees. In this instance nature, not history, has conspired to diminish diversity as a factor in the paleoanthropological consciousness; the result, however, is much the same.

But if you step back a little and look at the patterns of morphological distinction among primates that have a little more species diversity, you notice one thing very clearly: certainly as far as bones and teeth are concerned, closely related species (those belonging to the same genus - *Homo*, perhaps?) do not differ much. As I've noted, it's at the level of the genus that you pick up clearly recognizable differences. Most of the time, you will have to look pretty closely at the bones and teeth to discriminate consistently between members of two species belonging to the same genus—and even then you often cannot be sure. This in turn means that looking in the fossil record for infraspecific taxa—namely, the subspecies beloved of paleoanthropologists—is a totally futile pursuit. In contrast, however, when you find yourself comparing fossils that fall into two recognizably distinct "morphs," you can be pretty certain that you have (at least) two species in your sample.

This realization, born of years of studying the diversity of the lemurs (the "lower" primates of Madagascar), made conventional interpretations of species diversity in the human fossil record look a little odd to me. Oddest of all in the mid-1980s was the way in which most paleoanthropologists divided up the fossils representing the most recent half-million years or so of human evolution. Everything from this period (except for a few late *Homo erectus* stragglers such as those from Zhoukoudian) was classified in *Homo sapiens*. Yet there was among these fossils a very large amount of morphological variety, and this variety was well enough compartmentalized for at least three informal names to be in common use for different groups of them: Neanderthals (a.k.a. *Homo sapiens neanderthalensis*), "archaic *Homo sapiens*" (just about everything else that didn't happen to look like us), and "anatomically modern *Homo sapiens*." Well, if various groups of fossils are distinct enough to be identified by name, you can be pretty sure that you have at least as many species as you have names. I suggested this in a paper published in 1986; and, while I can hardly claim that my contribution revolutionized paleoanthropology, I think it was at least symptomatic of a trend that has recently gathered some steam. Specifically I urged that, at the very least, the Neanderthals be restored to separate species status as *Homo neanderthalensis*, Similarly, the Arago, Petralona, Bodo and Kabwe fossils should be classified together with others like them in their own species. If the Mauer jaw belongs to this group, as we can reasonably assume—if not prove, for lower jaws don't have a lot of diagnostic characters—we can call this one *Homo heidelbergensis*. Most emphatically of all, I stressed that our own living species, *Homo sapiens*, is as distinctive an entity as exists on the face of Earth, and should be dignified as such instead of being adulterated with every reasonably large-brained hominid fossil that happened to come along.

I pointed out also that the later part of the Pleistocene had been a period of extreme fluctuation in climate. Vegetation zones had moved north and south, up-mountain and down. Extensive woodlands had been fragmented by invading steppe or savannas, and had been rejoined as forests returned. Glaciers and harsh periglacial climates had made vast areas of northern Eurasia periodically uninhabitable by hominids, presumably spurring major migrations and causing local extinctions. Sea levels had risen and fallen, alternately creating islands and producing land bridges. Perhaps no period in the history of the globe had been more conducive to the emergence of new species and to competition between related species newly in contact—in other words, to evolutionary change. And the variety seen among later Pleistocene hominid fossils was, in fact, exactly the kind of thing one might expect to find under these conditions.

An origin of *Homo sapiens* through normal processes of allopatric speciation implies that, far from being a worldwide phenomenon, this origin was associated with a particular area of the world. And even as the multiregional continuity bandwagon gathered momentum an alternative scenario was beginning to be aired. Eventually becoming familiar as the "out-of-Africa" hypothesis of modern human origins, the basic notion was that *Homo sapiens* arose somewhere in Africa at a relatively recent date. Subsequently, mimicking *Homo erectus*, these ancestral modern people spread out from that continent to colonize all the habitable parts of the Old World—and eventually, of course, of the New World too. The crucial difference was that, while *Homo erectus* had dispersed from Africa into virgin territory, the early *Homo sapiens* had moved into regions already occupied by hominid relatives, which, necessarily, were displaced in the process. An early proponent of one version of this viewpoint was the University of Hamburg's Günter Bräuer, who in several papers published in the mid-1980s pointed out that, sparse as the relevant fossil record was, the earliest evidence for modern human anatomy came from eastern and southern Africa. Bräuer saw the migrant Africans as having displaced the Neanderthals in Europe, but he was unsure about what had happened in eastern Asia. And though others, such as Chris Stringer of the (then) British Museum (Natural History), were less reticent in generalizing the out-of-Africa scenario to embrace the Old World as a whole, what really catapulted it into the limelight was a series of genetic studies done in the laboratory of Allan Wilson, at the University of California, Berkeley.

Wilson and his colleagues had actually revived an approach originally initiated in the mid-1970s by the geneticists Masatoshi Nei and Arun Roychoudhury, but which had not attracted much attention at the time. Nei and Roychoudhury had looked at blood proteins in members of human populations Europe, Africa, and Asia. They had found that interpopulation differences were rather small compared to those found within populations, but that the net differences between Africans and both other groups were significantly larger than those between Europeans and Asians. This, they concluded, meant that Europeans and Asians shared a common ancestry more recently than either did with Africans, and they calculated that the split between Africans and Eurasians had taken place at about 115–120 kyr ago, while the European-Asian divergence was only about 55 kyr old. Wilson and colleagues took this approach farther, specifically by examining mitochondrial DNA (mtDNA).

Deoxyribonucleic acid is, of course, the long-stranded molecule that carries the genetic instructions from which each new individual is built. Most DNA resides in the nucleus of the cell, and in sexually reproducing species each individual's nuclear DNA is inherited in more or less equal proportions from each parent. However, a small amount of DNA is also contained within a distinct structure inside the cell known as the mitochondrion; what is particularly interesting about this mtDNA is that it is inherited only from the mother (whose ova are entire cells, whereas the father's sperm contains only nuclear DNA). Thus, while nuclear DNA is jumbled up in each new generation, mtDNA passes from mother to child pretty much unscathed—except for any mutations that may crop up along the way—in an unbroken sequence that stretches back to the ancestral "Eve". And in mtDNA mutations appear to accumulate at many times the rate typical of nuclear DNA, perhaps because they are not affected by natural selection in the way in which many changes in nuclear DNA are.

From the first, studies of human mtDNA showed a remarkable uniformity within the species. And since diversity is expected to increase with time, this uniformity suggested a relatively recent origin for *Homo sapiens*. By assuming an average rate of change of about three percent per million years, Wilson and his colleagues initially came up with a molecular time-to-origin for *Homo sapiens* of about 400 kyr. This would have placed the ancestress of modern humanity late in the time span of *Homo erectus*, though this was still hardly congenial to the multiregional continuity crowd, whose perspective stretched back over twice that far. It was, however, also at odds with what the fossil record tells us about the earliest appearance of modern human anatomy, and energetic debate ensued. Within a short time Wilson and his collaborators reduced the age of "Eve" to about 200 kyr, and since Wilson's premature death in 1990 his student Mark Stoneking has recalibrated this date to about 140-130 kyr—which fits well with the dates from Klasies River Mouth and elsewhere.

Molecular dates, which depend on a lot of assumptions, will always be argued over. More significantly, though, African mtDNA, just like Nei and Roychoudhury's proteins, turned out to show considerably more diversity—that is, a greater accumulation of mutations—than is found in the mtDNA of Europeans, Asians, and Australasians. Since mtDNA diversity is in some way a function of time, it follows that the African population has been evolving for the longest time since the genetic bottleneck at its origin; the less diverse populations split off more recently. "Eve", in other words, was African. Comparative analyses of the structure of mtDNA in these populations also initially suggested the same thing, although subsequent work has shown that those particular analyses were flawed. Nonetheless, the diversity data do seem significant, the rather homogeneous nature of human mtDNA (and thus a relatively recent origin for *Homo sapiens*) seems fairly well established, and, most significant of all, the mtDNA data do agree with what an admittedly less-than-perfect fossil record is also telling us.

The debate between the apostles of the multiregional and out-of-Africa scenarios was just beginning to attract the kind of media attention formerly associated with the Leakey-Johanson brouhaha, when new dating methods began to make the fossil picture of the origin of *Homo sapiens* more suggestive yet. The time gap between the effective ranges of radiocarbon dating and potassium-argon and similar techniques had left a very significant period of human evolution without any means of chronometric calibration. The 1980s saw the introduction or refinement of a number of new dating technologies that helped to fill this gap. Three of them, at least, deserve special mention here: electron spin resonance (ESR), thermoluminescence (TL), and uranium-series (U-series) dating.

Electron spin resonance dating springs from the observation that bombardment of crystalline materials by natural radiation leads to the trapping of free electrons in defects in the crystal lattice. The rate of trapping is determined by the level of background radiation. The energy of the trapped electrons can be measured and a date derived from the ratio between this figure and the trapping rate, which is in turn obtained from measurement of the background radiation (the "external dose") plus radiation from unstable isotopes absorbed by the fossil itself (the "internal dose"). The background rate can vary from one place to another, and even within the same site. So even though the electron spin resonance technique can be used directly on substances such as the dental enamel of fossils themselves (currently the material of choice) or on contemporaneously formed materials such as calcite, there must be enough of the initial archaeological or geological deposit left to allow accurate measurement of the background radiation. With older sites this isn't always the case, and in such instances accurate dates can't be obtained. The internal dose depends on the rate of uptake of radioactive isotopes by the fossil; this can't be measured directly, but the date can usually be bracketed by certain limiting assumptions. There are a number of additional complications, too, which narrow the range of potentially datable materials and deposits; but the reliability of dates obtained by electron spin resonance is rapidly improving, and the technique holds great promise for the future, as well as surprises in the present.

Thermoluminescence dating is based on similar principles to ESR, but the trapped electrons are measured in a different way. Again, the idea is to measure the number of electrons that have become trapped in the crystal lattice of a mineral. When the mineral was formed all of the traps were empty, but at that point they began to fill up at a regular rate, again determined by background radiation and other specifiable factors. Archaeologists, however, aren't interested in when a mineral was formed; they want to know when it was that humans used it. Heating and certain other processes empty the electron traps and thus reset the clock to zero; this is why flints burned in campfires have become popular objects for dating, as, for recent periods, have bits of pottery. Even exposure to sunlight can have reset some small objects; thus under certain conditions various kinds of artifact- or fossil-enclosing sediments can also be dated. The actual dating is done by heating the specimen once more, and measuring the intensity of light given off as the trapped electrons are released. If you know the background radiation that stimulated the filling of the traps, plus the sensitivity to radiation of the material itself (which can be determined experimentally), you can then calculate the date at which your flint was heated, or when the sand grains around your artifacts lay upon the surface.

Uranium-series dating relies on a different principle. Unstable uranium atoms decay at characteristic rates to various different daughter products. The daughter product most favored by archaeologists is Thorium-230 (^{230}Th), and the preferred materials for dating are freshwater-deposited limestones such as stalactites and travertines. Normally the radioactivity of ^{230}Th in an undisturbed material will be equivalent to that of the parent uranium in the sample. However, this equilibrium will not obtain in newly formed limestones. Since uranium is soluble in the water that deposits these limestones, whereas thorium is not, the newly forming stalactite will contain uranium but no thorium. At that point the ratio of thorium to uranium in the stalactite will be zero. As time passes, however, uranium will decay to thorium and this ratio will increase. The age of the stalactite can be estimated from the size of the enlarging ratio. Techniques for measuring the isotopes of uranium and thorium are improving, and along with them the accuracy of this method of dating. The particular attraction of U-series dating is that many archaeological sites are in caves in limestone regions. Datable travertines, flowstones, stalactites, and so forth, are, of course, common in such places. And, at least potentially, other calcium-containing structures such as bones, teeth, and mollusk shells are also datable using this technique.

These and other new dating methods have had a major impact on our

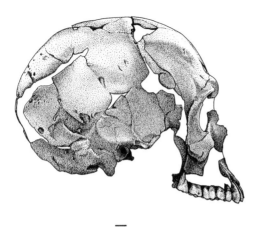
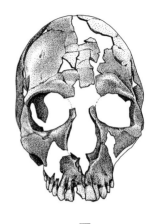

—   —

Side and front views of the ancient modern human cranium Qafzeh 9, from Jebel Qafzeh, Israel. Scales are 1 cm.   DM.

understanding of the emergence of *Homo sapiens*, and of the period during which *Homo sapiens* and *Homo neanderthalensis* coexisted. Uranium-series and ESR dates have, for example, combined to confirm the great antiquity (over 90 kyr) of the Middle Stone Age modern humans from Klasies River Mouth, in South Africa; they suggest also that the Border Cave moderns are in excess of 75 kyr old. But the real surprises have come from sites in the Mediterranean basin. Electron Spin Resonance dates on mammal teeth from the same levels as the human burials at Skhūl and Tabūn have come out at about 100 and 120 kyr, respectively. You'll recall that the hominids from the former are anatomically modern, or close enough to make no difference, while those from the latter are indisputably Neanderthal. Both dates are much more ancient than anyone had expected, and they confirm that Neanderthals and moderns coexisted in the Levant in some fashion at a very early date. The exactly nature of that coexistence remains unclear. It's possible that the two hominid species lived cheek-by-jowl (Tabūn is a leisurely five-minute stroll from Skhūl), though intuitively this seems unlikely; perhaps more plausible is that the two species occupied the area alternately, maybe as the local climate fluctuated. But whatever the case, coexistence was a long-term phenomenon, for Neanderthals lingered in the region undiluted until at least about 50 kyr ago, based on a very recent ESR date from Kebara. Thermoluminescence dates on burned flints from Jebel Qafzeh, where the human remains are modern by anybody's standards, confirm this lengthy overlap by revealing an antiquity of over 90 kyr. And, on the African side of the Mediterranean, the "modernish" skulls from Jebel Irhoud have been dated by ESR at well over 100 kyr.

This is very different from the traditional picture formed on the basis of western European evidence. There, the story is more clearly one of relatively abrupt replacement of the Neanderthals by modern types (although as later dates for Neanderthals come in from Iberia, and as recalibration of radiocarbon dates pushes back the dates of early modern occupation of Europe, clarity diminishes a little). What I find particularly interesting, though, is that at just about the time of the latest Neanderthal fossil (not, of course, the same thing as the last Neanderthal) known from the Levant, we find the first evidence of Upper Paleolithic industries in the region. In Europe such industries are associated with undisputed moderns; and although the very earliest Levantine Upper Paleolithic (from the 45 kyr site of Boker Tachtit, in the Negev desert) is not associated with human fossils, it's reasonable to believe that anatomically modern humans (who had, after all, been around in the area for 50 kyr) were responsible for it. Can we then say that Neanderthals disappeared from the area when *Homo sapiens* stopped simply looking modern and began to behave in a modern fashion, too? I'd personally bet that we can; but until we have more and more precisely dated archaeological sites and hominid fossils from this time and place, we won't be sure.

The subject of behavior is complicated by the fact that whereas in Europe Upper Paleolithic stone and bone tools were associated from the beginning with evidence of "creativity" in the form of engravings, sculpture, notation, musical instruments, and so forth, this was not the case in the Levant. What's more, the earliest Upper Paleolithic tools from Boker Tachtit, while fully Upper Paleolithic in concept, were made using techniques that had been current in the Middle Paleolithic. However, since anatomically modern humans had made Middle Paleolithic tools for the first 50 kyr of their existence, we probably shouldn't find this too surprising.

If the Neanderthals and moderns shared the world between them over an extended period of time, it's vanishingly improbable that there was no interaction between them. And, if interaction there was, what was its nature? One school of thought, in which the advocates of regional continuity figure prominently, finds evidence in variable morphologies of hybridization between Neanderthals and moderns. To them, the distinctive Neanderthal morphology was eventually "swamped" by incoming modern genes. A number of factors argue against this, however. One of these is that evidence of "hybrid" fossils (or fossils that can be interpreted as such) is poorest - indeed, as far as I can see, totally lacking—in just that region of the world where evidence for long-term cohabitation is best. Another is that, if the Neanderthals were a separate species from us—which the continuity people would deny, of course—significant interchange of genes would not have been possible (though, possibly, individuals might willingly or unwillingly have participated in attempts to hybridize). And yet another comes from observation of the appallingly nasty ways in which invading modern peoples have tended to treat each other—let alone other species—throughout recorded history. The idea of a gigantic blissful late Pleistocene love-in among morphologically-differentiated hominids simply defies every criterion of plausibility, however much we might wish to imagine otherwise. There are other reasons, too. If Lewis Binford is right about Neanderthal behavior, for instance, the incompatibility in behavioral systems between Neanderthals and moderns contemporaneous with them would make successful intermixing highly implausible. But suffice it for the present to note that both the fossil and the archaeological evidence needs to be passed through a powerful filter of perception before it's possible to swallow the conclusion either that Neanderthals are our forebears, or that those forebears somehow incorporated the Neanderthal gene pool into their own.

As the fossil record grows, and as dating is refined, we can expect the picture of the emergence of our own species to become clearer—or at least we can hope that it will. Meanwhile, we have to bear in mind that the emergence of *Homo sapiens* is the most recent major event in the evolution of our lineage, and that this fact has its downside as well as its advantages. Among the advantages is the rather large number of sites and fossils known from the period following about 100 kyr ago; but our close perspective also forces us to look at the picture in much finer grain than at any comparable earlier event in the human fossil record. Speciation is not a simple process, and what little we know about it doesn't predict much about what we should expect to see in terms of anatomical evidence. Testing alternatives is thus not easy at the best of times, and it is most difficult of all when we are working with a lot of widely scattered and frequently fragmentary fossils.

Perhaps this is not the most satisfactory of all notes on which to end a story; but as we arrive at the state of play today after looking at how our knowledge of our past as a species has developed over the last couple of hundred years, we have to bear in mind that this is a story that will never be ended as long as *Homo sapiens* persists on Earth. I don't pretend that what you have read so far in this book is objective history in the strict sense; it is simply the perception of one active practitioner of the science of paleoanthropology. But I do hope that it will have helped to produce some perspective on what we think we know today. As I said right at the beginning, what we think today depends very largely on what we thought yesterday. If the entire human fossil record were to be discovered tomorrow, and studied by experienced paleontologists who had developed their skills in the absence of preconceptions about human origins, I am pretty sure that (after the inevitable bout of intellectual indigestion) a range of interpretations would emerge that is very different from those on offer now.

With this caveat, then, let's go back to the basic fossil evidence of our own origins and emergence, bearing in mind our historical interpretive burden. We all want to know where we came from; and knowing where received wisdom colors our perceptions of our origins may help us to approach this question a bit more dispassionately.

17

Where Are We?

Given the wealth of interpretations available, it's a tall order to encapsulate the state of play in paleoanthropology today. But if you've read this far, you're already familiar with the most important constituents of the human fossil record and with the principal interpretations of them that have been made. What I can most usefully do by way of summary is, I think, to follow my own advice and to review the evidence for the past of our species by advancing from the simple to the complex: from a cladogram, to a phylogenetic tree, and finally to a brief scenario of our evolution. But first of all, we need to look at the most basic level of analysis of all: species diversity in the human fossil record. After all, before you can begin to work out the relationships between extinct species in the human family, you have to have a reliable idea of how many such species there are among the many hundreds of hominid fossils known.

We've already seen that this has not traditionally been anthropologists' strong suit, and that estimates of the number of extinct hominid species vary widely. There exists no recent minimalist appraisal of the number of australopithecine species, but I would guess that since the demise of the single species hypothesis not many (if any) paleoanthropologists would recognize fewer than three or perhaps four: *A. afarensis, A. africanus, A. robustus,* and (perhaps) *A. boisei.* In the genus *Homo,* by contrast, the minimalists would accept a mere two species: *H. habilis* and *H. sapiens.* In essence, this would reduce the number of species lying between us and our earliest known bipedal ancestors to only two: a ludicrously inadequate figure given what we know about the amounts of bony anatomical difference that are typically found among closely related living species.

A more mainstream assessment would add to the genus *Homo* at least one more species, *erectus*, plus the informal categories of "archaic *Homo sapiens*" and the Neanderthals. My own preference, though—and, I think, that of a growing number of colleagues—would be to recognize at least three genera in the human family, embracing perhaps a dozen species. In this view, the genus *Australopithecus* accommodates only *A. afarensis* and *A. africanus*, the "robust" group being placed in its own genus *Paranthropus*. (I should probably remark at this point that if *afarensis* is truly a "stem" species that gave rise to *africanus* on the one hand and to the species of *Paranthropus* on the other, it would, strictly speaking, require its own separate generic designation. In practice, however, paleoanthropology is not quite ready to contemplate this possibility; and since the reidentification of the growing number of fossils usually allocated to *A. afarensis* promises to be one of the major debating points in the field over the coming years, any concrete suggestion to this effect would anyway be more than a little premature.) Within the genus *Paranthropus*, at least three species are clearly diagnosable as distinct: *P. robustus*, *P. boisei* and *P. aethiopicus*. Some paleoanthropologists would even squeeze in a fourth, *P. crassidens*, in the probably accurate belief that the robusts from Kromdraai and Swartkrans represent different species. Whether *aethiopicus* is the appropriate name for the species containing the "Black Skull" is a debatable point, but the current conspiracy of silence on that matter is clearly desirable and one hopes that it will be maintained.

The plot thickens further as we approach the genus *Homo*, largely because there is no clearly evident reason why it should include such forms as the gracile fossils from the lower levels of Olduvai. As we've seen, the Olduvai remains were initially allocated to our genus because of presumed stone tool making, a fractional increase in brain size relative to *Australopithecus*, and, perhaps above all, because of Louis Leakey's desire to justify his longstanding belief in the ancientness of *Homo*. In no other group of mammals would such considerations be accepted as justification for grouping species as unlike as *habilis* and *sapiens* in the same genus; and even though the taxonomic game seems doomed forever to be played in Hominidae by its own distinctive set of rules, the sheer primitiveness of the postcranial skeleton of the recently discovered Olduvai Hominid 62 should be sufficient to make paleoanthropologists think again about genus-level diversity within Hominidae. Alas, however, they almost certainly won't, so we must continue to live with our curiously inflated concept of the genus *Homo*. This being the case, how many species of it do we know?

For the moment there is no well-argued alternative to Bernard Wood's conclusion that all of the gracile Olduvai fossils are allocable to the species *Homo habilis*, along with such specimens as the ER 1813 cranium from Koobi Fora and Stw 53 from Sterkfontein. So, provisionally, it's reasonable to accept it. It's even more reasonable to concur with Wood that specimens like the Koobi Fora cranium ER 1470 are quite distinctive and should be placed in the separate species *Homo rudolfensis*. Hard on the heels of these species comes *Homo ergaster*, which, as the "Turkana Boy" attests, is postcranially as well as cranially altogether more modern. Fred Grine has recently reconstructed the Swartkrans SK 847 partial face, long considered to be a close match for the Koobi Fora *Homo ergaster* specimen ER 3733, and has been able to demonstrate that certain differences exist between the two. Whether or not this means that a separate species related to East African *ergaster* lived in South Africa at about the same time, it's a bit early to say for sure. On the other hand, what is very clear, at least to me, is that *H. ergaster* is not the same thing as the classic *Homo erectus* from eastern Asia. A related species, yes; but the African form is more primitive than the Asian one, and it deserves recognition as a distinct species. Later in time, we begin to encounter distinctive early human types that have long been distinguished by informal names, and that are due by now for formal recognition. "Archaic *Homo sapiens*," as exemplified by such specimens as Bodo and Arago, deserves its own specific epithet, probably *Homo heidelbergensis*, while there can be no doubt at all as to the correct species name for the Neanderthals: *Homo neanderthalensis*. All humans living today, and all fossil humans who resembled us, are *Homo sapiens*.

This brief summary of species recognizable within the genus *Homo* excludes a number of individual fossils that are suggestively distinctive, but which by themselves don't offer us the weight of evidence necessary to establish new species. Bob Martin recently estimated (in my view, generously) that the fossil record provides us with evidence for only about three percent of all the primate species that have ever existed; and it's pretty evident that even if we accept six species within the inflated genus *Homo* we are still liberally underestimating the actual species diversity of hominids over the past two to two and a half million years. Nonetheless, if we resist the traditional urge to expand the species *Homo sapiens* beyond all biologically plausible limits, we can still obtain a basis for proceeding to the next level of analysis: our cladogram.

However many species of extinct hominids you accept, the relationships among them are, and will continue to be, the subject of vigorous debate. It would be pointless to pretend that I have any hope of resolving any of the current uncertainties here; the accompanying V-diagram is simply a way of expressing one reasonably plausible set of relationships among our dozen species. Like all cladograms, this diagram does nothing more than express relative closeness of relationship. It says nothing about exactly what kinds of relationships are at issue: for example, whether they are between an ancestor and its direct descendant or simply among descendants of the

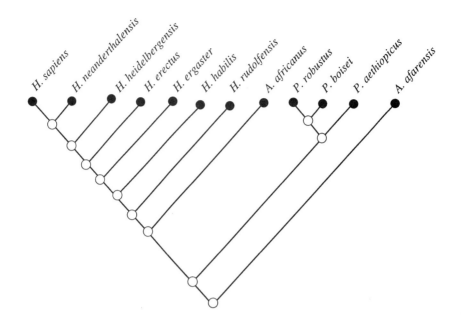

Cladogram showing a suggested scheme of relationships among the species belonging to the human family. If correct, this cladogram illustrates well the problem raised by including A. africanus *and* A. afarensis *in the same genus.* DS.

same common ancestor. Neither does it say anything about time. Such additional details are the province of the evolutionary tree, an example of which is seen on page 233. Of course, trees are more complex statements than are cladograms, and a single cladogram may be compatible with several different trees. What's more, if pressed I would have to admit that, even if all the fossils comprising the human fossil record are necessarily someone's descendants, probably rather few of them represent the direct ancestors of any later fossils in that record. The tree here is only one of many possible, and there are many points of potential dispute in it. However, taking ancestry and descent in a broad sense, it's a reasonably justifiable reflection of the pattern of hominid evolution—as it's sampled today by known fossils.

Evolutionary trees are visual devices, and as such speak for themselves with little need for additional narrative explanation. The only point about our tree here that really needs special stress is the weakness of the link between the remarkably primitive *Homo habilis* and the fractionally later but substantially more advanced *Homo ergaster*. Where narrative comes into its own, however, is in the fleshing-out of the tree into a full-fledged evolutionary scenario. Scenarios are, of course, the most interesting kind of evolutionary statement, involving as they do questions of adaptation and environment as well as the basics of time and relationship. On the other hand, as I've noted already, they are generally so complex that weighing alternative scenarios against each other is usually more a test of the storyteller's art than of the paleoanthropologist's science. Strictly speaking, it is only at the level of the cladogram that rigorous comparison between competing hypotheses is possible, although for most tastes trees are comparable enough as long as you know the cladograms that gave birth to them. Scenarios are quite another kettle of fish: in this arena you pays your money, and you takes your choice. Most of the ingredients for a wide variety of full-scale scenarios of human evolution are already contained in earlier chapters of this book, and everyone will wish to mix those ingredients personally. But let's complete our review by presenting one abbreviated scenario that attempts to round out the picture as presented so far.

This story starts with a bang, if for no better reason than there are no fossils that document humankind's initial and presumably painful descent from the trees. With the exception of a handful of fragments that do no more than hint that early hominids were in some way around between 5 and 6 myr ago, the first evidence of our precursors on the world—for which, read African—stage comes with *Australopithecus afarensis*.* This eastern African species is clearly diagnosable from fossils only at well under 4

---

*As this book goes to press, the discovery of fossils allocated to a new species of *Australopithecus, A. ramidus,* has been announced by Tim White and colleagues. Found at a site known as Aramis, in Ethiopia's Middle Awash region, and dated to about 4.4 myr ago, these fossils consist of a variety of isolated teeth, some fragments of cranium, and some arm bones. As would be expected of a human species antecedent to *A. afarensis,* the teeth seem to be more primitive in certain respects, for instance in canine size and lower premolar morphology. The cranial and arm bones are not very informative, although the former are said to be consistent with upright posture. Perhaps most telling is that the sediments which yielded these fossils were reportedly laid down in a fairly densely wooded environment; if this is the case, then these early hominids were not necessarily (or even at all) savanna dwellers. This might not be surprising in view of the recent demonstration by Dick Rayner and colleagues that woodland conditions also reigned at Makapansgat at about 3 myr ago; but it has sparked speculation (probably premature, especially given that bipedalism in *A. ramidus* has yet to be demonstrated) that the adoption of upright locomotion by early humans was not related to the expansion of savanna conditions at the expense of forests. The causes of upright bipedalism (was it primarily a locomotor adaptation, or a physiological accommodation, or something else? Did it indeed represent a response to environmental change?) are clearly matters poised for much future debate.

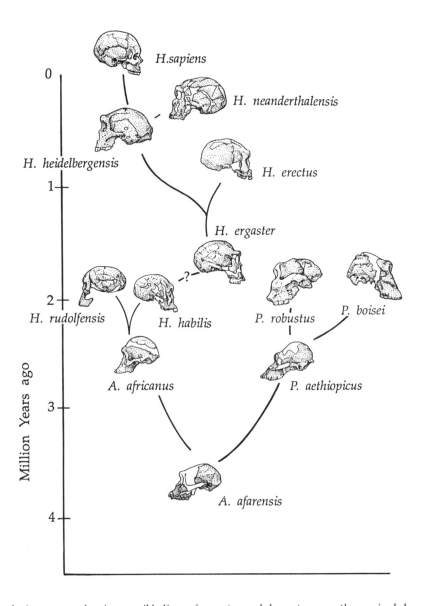

Evolutionary tree showing possible lines of ancestry and descent among the species belonging to the human family. From I. Tattersall, The Human Odyssey, *Prentice Hall, 1993.*

DS.

myr ago, by which time an upright bipedal body structure was plainly established—though it was combined with rather apelike proportions of the head. Nobody would dispute that the adoption among hominoids of upright bipedalism represented an adaptive breakthrough; and analogous episodes in the long story of life on Earth suggest that radical innovations have often been followed by rapid evolutionary diversification within the group involved. Nonetheless, right now there is not much evidence in this instance to indicate any very substantial diversification. Perhaps, then, this vaunted "breakthrough" appears as such only in the retrospect of an ego-centric species!

What there is good circumstantial evidence for, however, is that the development of upright bipedalism by the ancestral hominid was intimately tied up with a change in climate and environment,* that it somehow represented a response to the shrinking and fragmentation of formerly extensive forests in the part of Africa lying in the rainshadow to the east and south of the great domed-up Rift Valley. As the forests became more discontinuous, eventually becoming confined to river margins and lake basins, the protohominids were forced to travel between wooded remnants that in themselves did not provide sufficient sustenance year-round for foraging groups of such large-bodied primates. Travel in open country provided physiological challenges that had not confronted the protohominids' forest-living precursors, but the adoption of full-time bipedalism on the savanna was probably an efficient way of getting around for a hominoid that already had a penchant for carrying its trunk upright. And it also, as we've seen, provided advantages in brain and body cooling and in reducing water dependence. Loss of speed was presumably not a critical factor, since open-country predators would have had the advantage over even the fleetest of hominoids; trunk uprightness, on the other hand, would have helped preserve the defensively useful climbing abilities inferred from the limb proportions and from the hand and foot structure of *A. afarensis*.

Bipedal locomotion doubtless had other advantages, too, out there on the open savanna. Thus both the distance to the horizon and the visibility of potential predators—especially in long grass—are vastly increased by raising eye level, while a vertical silhouette is less attractive to a carnivore than is a horizontal one. What's more, the savanna was never simply a barrier to be crossed in the most efficient and least hazardous manner; it also represented a whole new range of food resources for creatures with the wit to exploit them. Such resources included the roots and rhizomes of shrubs and grasses, and—most fatefully—the remains of dead animals, an incomparable source of fats and proteins. Descended ourselves from the most expert of hunters, we are prone to think of the scavenging of animal carcasses as a pretty humble and unsophisticated kind of occupation; but in reality this activity demands a wide variety of skills. Much has been made, for example, of the importance of the creation of "mental maps" of where food resources are to be found in the development of mankind's vaunted

*See footnote, page 233.

cognitive skills. The fact is, however, that any primate you can name, certainly any fruit-eater, possesses the capacity to patrol its territory and to predict where certain resources are to be found in different seasons. What is different about scavengers on the savannas is in reality the lack of maps: lion kills are scattered about much less predictably in time and space than are fruiting trees, which can be relied upon to perform on a more or less regular schedule. To find carcasses out in the grasslands, you have to learn to read indirect signs: vultures wheeling in the air above; movements of herd animals. Of course, before the invention of stone tools there was a limit to the extent to which it was possible to exploit the kills of other animals; but the potential was there, and the savanna also provided opportunities to hunt smaller and more helpless forms of life, as chimpanzees have been observed to do in woodland settings.

It is hazardous enough for slow-moving creatures of Lucy's size simply to be out on the savanna. Even if getting from one patch of forest to another was what enticed such primates out there in the first place, it seems probable that scavenging and possibly root-digging were what kept them there. This isn't to say that early hominids turned their backs on their arboreal past in a single final gesture—clearly, they didn't. Probably for nigh on two million years or more australopithecines maintained a "have your cake and eat it" adaptation, combining efficient bipedal locomotion (and freed-up hands) on the ground with the grasping abilities, long arms, and narrow shoulders of reasonably accomplished arborealists. Not only would this particular combination of abilities have allowed these hominoids to shelter with ease in the (comparative) safety of trees, but it would have suited them well for such highly specialized activities as stealing leopard kills lodged in trees, while also taking advantage of terrestrial food resources. What is most certain of all about the basic australopithecine strategy is its long-term success, compromise between two lifeways though it may in origin have been. For it seems to have endured essentially unchanged for the entire first half of documented hominid existence.

What of the parting of ways between the "robust" australopithecines and the other early hominids, and of the divergences among the robusts themselves? This question doesn't demand much elaboration here because it can, I think, be fairly satisfyingly answered in terms of specialized diet and environment, on the one hand, and of the common tendency of lineages to split (especially in changing environments), on the the other. What's more difficult for the scenario writer to explain is why the invention of stone tools seems to have had so few immediate consequences (apart, presumably, from increased scavenging efficiency) in the lineage that led ultimately to us.

The very earliest stone tools turn up at about 2.5 myr ago, and this is also about the earliest date we have for *Homo rudolfensis*, if (maybe a big if) a newly discovered jaw from Chiwondo in Malawi belongs to that species. *Homo rudolfensis* was endowed with a brain larger than that of any australopithecine and seems to have been a little more modern in its limb morphology; however, it also had some specializations which make it less likely than the smaller-brained and more primitive-limbed *Homo habilis* to have given rise to later humans. Exactly who made the oldest tools known from eastern Africa we don't know; all we can say is that *Homo rudolfensis* was (probably) around somewhere at the time. But if *Homo habilis* was, as seems virtually certain, the toolmaker of Bed I at Olduvai Gorge, this new capacity (which Toth and Schick's bonobo experiments indicate did indeed represent a major cognitive advance) was hardly accompanied by any revolution in anatomy or lifestyle. This leap forward was made by a hominid that didn't look very different from—or, as far as we can tell, behave very differently from—its australopithecine predecessors. And this brings us to one of the themes that has consistently marked the course of human evolution: the decoupling of technological change from anatomical change.

In a way, this lack of association seems counterintuitive, for it is both easy and satisfying to "explain" any given technological (and putatively cognitive) advancement by invoking a new kind of hominid, brandishing a new kind of structure—a hand capable of precise manipulation, for example, or a brain able to make new associations. But if you think about it for a moment, you realize that there's no plausible way for things to happen like that. Species just don't saltate into vastly different other species. Any behavioral innovation—indeed, any innovation at all—has to take place *within* a species. There's simply no place else for it to happen. Innovations, whether genetic or behavioral, ultimately originate with individuals; and every individual has to belong to a preexisting species.

Perhaps this is where the famous incompleteness of the fossil record comes in most importantly; we just can't look at our prehistory in fine enough grain to detect all the numerous subtle changes that must have combined to give us the grosser effects that we're able in our telescoped retrospect to observe. Certainly, as we approach the present and the record gets denser, events of the apparently clear-cut kind that occurred in earlier times mysteriously begin to disappear. For example, although we may well debate the details, we have little trouble in principle with ideas such as that *A. afarensis* gave rise to a robust lineage on the one hand and a gracile one on the other. How *Homo sapiens* emerged from its precursor form is, however, a totally different kettle of fish—and a much more complicated one, as the later history of *A. afarensis* would doubtless also appear if we were able to observe it in greater detail. I have little doubt that the entire long story of human evolution is littered with minor biological changes—speciations—that, given the coarseness of the record stand out only in the aggregate.

On one level this may be less true of technological (or cognitive) novelties than of biological ones: after all, the difference between making stone tools—however crude—and not making them is certainly of night-and-day magnitude. And the evidence we have that suggests, for example, that early hominids carried lumps of stone around in anticipation of needing them as raw material for utensils certainly tells us something about these beings that we couldn't otherwise know. Nonetheless, the ways in which technological innovations affect lifestyles may be quite subtle, and it is perhaps in this context that we can best understand how it was that small-brained, primitive-bodied early hominids began to make stone tools yet continued to live and look pretty much as they already had for millions of years. For, while from a technological viewpoint the invention of stoneworking looks like a truly momentous innovation, at the time it appears simply to have enabled hominids to do what they had always done, just a little bit better— as, indeed, new generations of technology have continued to do ever since, each one building on the one that preceded it, and each ultimately tending to persist at least for a while alongside its successor.

It's not even clear that the arrival of modern body proportions heralded any major lifestyle changes among hominids; after all, the first *Homo ergaster* made tools that were more or less indistinguishable from those of *Homo habilis*, and there is not much evidence for other behavioral developments, either. This isn't, of course, to deny that, uprightness apart, the arrival of *Homo ergaster* marks the single most important transformation below the neck in the entire course of human evolution. Clearly it was, and equally clearly it was not something that was achieved in a single fell swoop. Beyond elegantly demonstrating the essentially modern and highly efficient striding structure of the Turkana Boy's skeleton, Alan Walker and his colleagues have pointed to a whole suite of characteristics that indicate very specific adaptation to an environment of high radiant heat stress. This kind of accommodation to the environment is something which the history of the differentiation of *Homo sapiens* (with populations as distinctive as Dinkas and Eskimos) shows need not take too long: certainly well under 100 kyr. It's less obvious, however, that a radical shift in body design such as that exemplified between Olduvai Hominid 62 (1.8 myr old) and the Turkana Boy (1.6 myr old) is something that one might expect to happen (in evolutionary terms) overnight. But unless the known fossil record is substantially biased, it's nonetheless apparent that the acquisition of essentially modern human body build and proportions was a relatively rapid process. The acceleration of this process was probably due to strong competition between differentiating hominid populations (species?) as they strove for survival in increasingly severe environmental conditions.

Following the emergence of the "package" of modern human postcranial adaptations it's not hard, at least in retrospect, to see why it became so quickly established. What may be more difficult to explain, however, is the enlargement in brain size that apparently accompanied it. The problem is that the brain is a very delicate and energy-hungry instrument that accounts for a substantial proportion of the body's energy requirements and that demands efficient cooling. Any enlargement simply exacerbates these needs; and where it occurs one might reasonably expect to find some kind of countervailing advantage accompanying it. What might that advantage have been?

One possibility is that at least much of the increase in brain size seen in *Homo ergaster* compared, say, to *Homo habilis* may have been due simply to an increase in body size: the more mass there is to control, the bigger the controlling organ needs to be. And increased body size was in itself presumably advantageous, for example, in reducing vulnerability to predators. There are mathematical ways of determining whether an observed brain size exceeds the "expected" value for a given size of body; the problem is that the expected value will differ depending on which group of other species you base your expectations on. But even on the most conservative assumptions it does seem that *Homo ergaster* showed some increase in relative brain size compared to earlier hominids, in which case we need to look elsewhere than simply to body size if we are to explain this modest brain enlargement. Unfortunately, the usual suspects don't help much. The archaeological record doesn't indicate that, at least to begin with, *H. ergaster* was a significantly more efficient exploiter of the environment than *Homo habilis* had been, and thus that the smarter species was able to increase its energy budgets; moreover, anatomical studies don't suggest that the Turkana Boy had been a great communicator, either.

Language, a uniquely efficient form of communication, has often been looked at as a key to brain size increase and the improvement of human intelligence and communication over evolutionary time. Various landmarks of the external surface of the brain, for example, have for years been minutely scrutinized by paleoneurologists for evidence of language capability. Unfortunately, it turns out that this capability is much less highly localized in the brain than many had supposed, which sets us back somewhat in the study of brain casts. Examination of the Turkana Boy's skeleton, however, opened up a new and unexpected avenue in unraveling the related capacity for speech. Ann MacLarnon, of London's Roehampton Institute, noticed that in this individual's spinal column the space for the part of the spinal cord that controls the chest muscles was much smaller than it is in modern humans. What could this mean? The most convincing explanation is that the Turkana Boy possessed less exact muscular control than we do of the walls of the thoracic cage. This space contains the lungs, which are the ultimate source of the vibrating column of air that the upper respiratory tract manipulates to produce speech. Imprecise control of the abdominal wall might well thus correlate with a lack of speech capability. Add to this that the flexion of the cranial base in *Homo ergaster* is not impressive enough to suggest a very significantly lowered larynx (though it may reflect an adaptation to extremely arid climes involving moisturization of inhaled air), and the conclusion that this species did not possess speech as we know it today seems unavoidable. Gestural and vocal communication skills were nonetheless presumably improving, perhaps as symbolic values were attached to a greater range of signs and sounds (and maybe more important, with additive effects in the case of strings of sounds). There is, of course, little direct confirmation of this to be found in the archaeological record, although it is difficult at this point to imagine exactly what form such confirmation might take.

This leaves us with few putative advantages to accompany the enlarged brain of *Homo ergaster*, and although it may simply reflect a lack of imagination, in certain ways I find this rather comforting. For, I have to admit, I don't belong to the school of thought which holds that species have to be exquisitely fine-tuned to their environments. Indeed, most widespread species manage to straddle several different environments with at best only minor variations in morphology, although these variations do, of course, provide the basis for evolutionary change. They also—since a structure has to exist before it can fulfil a function—give new capacities the opportunity to strut their stuff when the occasion arises. This was certainly the case with *Homo ergaster*, which was responsible for the next innovation in the technological record: the invention of Acheulean stone tools, which first turn up around 1.5 myr ago—about coincident with the first (putative) intimations of the control of fire.

You'll recall that the handaxe and cleaver, the typical Acheulean implements, were the first tools to be made to a set and regular pattern: to a "mental template" which existed in the mind of the maker. Oldowan toolmakers—as many Acheuleans also continued to do—had, in contrast, been most interested in producing sharp stone flakes that they could use as cutters and scrapers. They were thoroughly aware of the principles of striking stone at the correct angle for producing such flakes, but they were not concerned to produce tools of particular shapes. The Acheuleans were, however—and they came to revel in it. Identical handaxes litter many localities in almost unimaginable abundance, and at sites such as Tanzania's Isimila they may be of extraordinary size—in some cases too heavy to lift with one hand. Apart from showing us that the tool-makers were far stronger than any humans are today—for the shaping of large stone cores of this kind demands prodigious physical power—such tools as those from Isimila maybe even show a sense of humor, or at least a sense of pure design: some of these tools were simply too large to use, certainly with any deftness. Such caprices apart, however, the Acheulean handaxe has been aptly described at the "Swiss Army Knife of the Paleolithic," subserving a diversity of functions such as cutting, hacking, scraping, and digging. And its manufacture endured for well over a million years.

There's no doubt in my mind that with the invention of sophisticated bifacially flaked tools such as handaxes we are witnessing yet another major cognitive leap on the part of mankind. But, once again, this development was apparently independent of any other major innovation, biological or even cultural. For example, even once the Acheulean was well established, archaeologists are reluctant to conclude that *Homo ergaster* or its later relative *Homo erectus* ever systematically hunted large animals. And, despite fire's manifold modern mystical meanings, its domestication—if indeed it occurred at the beginning of the Acheulean—doesn't appear initially to have had a broad impact, either. The stone tools themselves apart, the leavings of Acheulean hominids didn't differ a lot from those of their predecessors, though Acheulean handaxes were excellent all-purpose tools for butchering animal carcasses. Once again, then, we find the persistence of a general lifeway even in the presence of technological refinements that one might have expected to do more than merely permit it to be carried out more efficiently, and maybe more safely. It's fair to remark, however, that the advent of such cognitive innovations as "mental templates" might well have changed the rules of the natural selection game by putting a premium on abstract intellectual skills, and might thus have deeply influenced future biological and cultural developments.

This may in fact be the key to the obvious question that much of the preceding discussion raises: if general lifeways changed so little prior to the advent of *Homo sapiens*, why do we see so much physical change—and particularly brain enlargement—in the course of human evolution? One possibility is that even relatively small technological advances—changes simply in the way in which the manipulability of the world was viewed by hominids—may have handed an evolutionary advantage to those individuals capable of grasping and exploiting them. This would quickly have affected the composition of the small groups and populations in which early humans lived, and in their turn the rapidly fluctuating climates—and geographies—of the Pleistocene would have promoted a rapid process of sorting among those populations. It will ultimately be for the archaeologists to provide the evidence to test at least the first of these possibilities: cognitive change reflects itself most directly in behavior, and fossil behavior is the province of archaeology. It will, though, require the analysis of the archaeological record in much finer grain than has typically been the case so far. More recent advances in Paleolithic archaeology have focused less on the finer reading of the actual record itself than on the potential deficiencies of earlier work (were those features at Terra Amata—or Olduvai—really structures? Is there really any evidence for Neanderthal "bear cults"?). Unhappily, this is inevitable, for the act of excavation destroys the record even as it exposes it, and meticulous documentation of what has been removed is a relatively new development in archaeology. Some of the most significant sites were dug during an era when archaeology was extremely low on its learning curve, and most of the information they contained is irretrievably lost. It's tragic, if inevitable, that at a time when archaeologists are finally beginning to understand how to deal with all the myriad complexities of their subject, many of the most important sites have been lost forever.

It is likely, then, that we are simply not in a position to discern all the important events in cognitive development that accompanied human physical evolution—and that presumably fed back into it. Nonetheless, lack of change in the archaeological record is clearly not simply an artifact of our limited perception. Thus, following the invention about 1.5 myr ago of handaxes and related bifacial implements such as picks and cleavers, we have to wait another 0.5 myr before we encounter any notable advance in stoneworking technology. And even here, change expressed itself simply as a refinement of the basic handaxe-making technique. At about 1.0 myr ago, handaxes began to be made thinner, using a technique known as "platform preparation," in which the axe's edge was initially made less oblique, to provide a surface at which more force could be directed. This advance was associated with at least occasional use of "soft hammers," made of gentler organic materials—bone, wood, antler—rather than of brittle stone. Who was responsible for this invention? We don't know; associated hominid fossils don't exist. But we can be pretty sure that these early hominids had some powers of abstract reasoning, for as Kathy Schick and Nick Toth point out, platform preparation is not an intuitively obvious procedure. Which raises a question we've so far avoided: could these advanced Acheuleans be described as human? In a strict legalistic sense, I suppose they must be considered *ex officio* humans, as members of the genus *Homo*. But that's not to say that we would intuitively recognize them as such if we were to encounter a group of them while out for a stroll on the savanna. In the absence of an agreed functional definition to tell us what is human and what is not, everyone has to make up his or her own mind; what is certain, however, is that even the latest Acheuleans were far from *fully* human as we are today.

The period around 1 myr ago was an eventful one in human evolution. For in addition to technological advances, it also witnessed the first well-documented emigration of hominids out of Africa, to populate wide stretches of Asia and perhaps of Europe, too. Actually, there are hints both from archaeology and from fossils that humans may have left Africa as much as 1.5 to 2 million years ago; but each such occurrence is individually questionable, especially as to dating, and certainty only begins at about a million years ago—when, for example, we begin to find evidence for *Homo erectus* in Asia*. Interestingly, stone tools from this period—and later—in Asia tend to be rather miserable things, although as we've seen there may be special reasons for this. Nonetheless, from a technological perspective this spread of hominids is highly suggestive, since it involved the occupation of temperate areas of the globe that required survival in conditions very different from the tropics. Again, though, it's hard at present to read the archaeological record in fine enough detail to know exactly how strategies evolved to cope.

Approaching the end of the Acheulean's heyday brings us to the period in which hominid brain sizes were inching up close to the modern average, which is the principal reason why most hominid fossils of the last half-million years or so have been described as "archaic" forms of *Homo sapiens*. Yet hominids of this period looked very different from us from the neck up, with foreheads retreating behind large brow ridges, and large faces with biggish teeth (compared to *Homo sapiens*), hafted in front of the braincase rather than beneath it. We have no good associated skeletons (at least that have been described) to tell us exactly what these hominids were like below the neck, but those bones that we do have suggest that their bodies were built much like ours, only a good deal more powerfully. At least one highly distinct species of this kind is known worldwide; this is *Homo heidelbergensis*, represented by fossils from sites as far apart as Kabwe in Zambia, Bodo in Ethiopia, Arago in France, Petralona in Greece, and (probably) Dali in China. Dating of most of these finds is imprecise, but in the same general time range of 400–200 kyr ago we also find (in addition to late-surviving

*Carl Swisher and colleagues at the Institute of Human Origins, Berkeley and in Indonesia have reported obtaining dates of about 1.8 and 1.6 myr for two hominid fossils from Java. Neither specimen is particularly diagnostic—the classic *Homo erectus* fossils remain to be dated definitively—but this new dating lends strong support to the idea that hominids first left Africa very early—and certainly well prior to the invention of handaxes. This may explain the virtual absence of handaxes at eastern Asian sites, and suggests that successive waves of new hominid species left Africa at various times.

H. erectus) fossils such as those from Ndutu and Steinheim, whose morphology does not fit comfortably within *H. heidelbergensis*. The actual diversity of hominid species at this stage of our prehistory is thus not entirely clear.

Some of these new kinds of humans (especially outside Africa) are associated with remarkably rudimentary stoneworking technologies, but in the 200 kyr time range we begin to encounter a much more sophisticated way of working stone that involves "prepared cores." In this technique a lump of stone was fashioned in such a way that a single final blow (usually with a soft hammer) would detach a "flake" that represented a quasi-finished tool. This was a considerable advance, for it ensured the production of a long, continuous cutting edge around almost the entire periphery of the tool. It is in this period, too, that we first begin to see the regular appearance of hearths—organized emplacements for campfires—in archaeological sites, as well as of evidence for domestic structures such as huts. Interestingly, we also begin to pick up evidence for the intentional defleshing of human remains, as evidenced, for example, by stone tool cut-marks on the forehead and within the orbit of the *Homo heidelbergensis* cranium from Bodo. Whether this scalping indicates cannibalism (which seems a little unlikely) or some other ritual behavior is anybody's guess. From this point on, too, stone tools began to decrease in size, although some of the older forms were preserved; for example, small handaxes were made on flakes, rather than on large cores broken off much larger boulders as was the practice in Acheulean times. Other flake tools might have been hafted onto handles or spear ends.

The apogee of toolmaking of this type was achieved by the Neanderthals, whose beautifully crafted Mousterian tools came in a large variety of standardized forms. Neanderthal fossils begin to show up in the record at some point between about 200 and 150 kyr ago (depending on how you interpret some rather fragmentary evidence), although some Neanderthal-like features have been detected in *Homo heidelbergensis*-like fossils dating from as much as 300 kyr ago. The Neanderthals, particularly the later "classic" forms of Europe, had brains that were fully as large as our own, although they were enclosed in a braincase of rather different shape, and nestled behind, rather than above, large faces of quite distinctive morphology. Although there has recently been some debate over this, it's clear that, at least occasionally, Neanderthals buried their dead. They also on occasion provided long-term support to disabled members of the social group, and may have decorated their bodies with ochre, although except for a very few traces in the Châtelperronian they do not appear to have decorated objects. At most Neanderthal sites there is little evidence of the deliberate organization of the domestic space, although Kebara may be an exception to this.

If Binford is correct, classic Neanderthal males and females led virtually separate lives and fulfilled entirely distinct economic roles; in any event, they did not leave behind them the evidence of the sophisticated hunting (and fishing) of the wide variety of animals that formed the prey of later, fully modern people.

In Europe the Neanderthals were replaced relatively abruptly by moderns, even if this process was not quite the simple one- or two-wave invasion sometimes envisioned by earlier paleoanthropologists. Nonetheless, replacement seems to have been the order of the day, rather than a commingling of modern and Neanderthal stocks. In the Levant, however, the situation was less clear-cut, Neanderthals and anatomically modern people coexisting or alternating in occupation of that region for upwards of 50 kyr. *Homo sapiens* itself seems most likely to have arisen in Africa, whence it migrated to replace more archaic peoples in (ultimately) all inhabitable regions of the world. The date of this origin was certainly in excess of 100 kyr ago, though by exactly how much remains a subject for speculation. The coexistence between Neanderthals and moderns in the western Mediterranean region is perhaps made less remarkable by the fact that the two shared virtually identical stone tool technologies—and presumably economic strategies—throughout their time of coexistence; it was only when people who were "anatomically modern" became associated with more "modern" behavior patterns as revealed by the archaeological record that cohabitation in the Levant finally ceased, and *Homo sapiens* took over definitively.

In Europe, by contrast, anatomically modern people brought modern behavior patterns with them, presumably accounting thereby for the rather rapid disappearance of the Neanderthals, who were gone by a little over 30 kyr ago. The Aurignacians, the first modern culture group in Europe, left immediate evidence of art, notation, symbolism, music, sophisticated manipulation of materials, a restless spirit of innovation, and all of those basic behavioral elements that we associate with ourselves. For the first time, we can be sure that people possessed articulate speech, whereas there is no way in which we can be certain of this in the case of any earlier group. Once again, then, there was a discontinuity between anatomical and behavioral innovation: people began to behave in a modern fashion only well after they had assumed modern anatomy, just as (presumably) *Homo ergaster* had invented the Acheulean only well after making its physical debut. Such decoupling is thus part of a well-established pattern in human evolution, rather than, as is often assumed, a special pattern to be specially explained. Less is known of the arrival of modern humans and the disappearance of earlier types in other parts of the world, although it is becoming evident that modern behaviors were being exhibited elsewhere some time before they were introduced to Europe. In Australia, for instance, some rock art has now been dated to over 40 kyr ago, and it is evident that considerable navigational skills had already been acquired by about 50 kyr ago. For the early colonists of Australasia, who we now know had arrived there by that time, had to cross at least fifty miles of open ocean.

Once modern behavior patterns had become established, the pace of technological change began to pick up on average, even though it appears that many populations were so well in harmony with their environments that their lifeways changed rather little over the last couple of dozen millennia. Again, this is a pattern that we see around us in our own society, with different "generations" of technology existing alongside each other even as the pace of innovation roars ahead. Around the world, societies have existed within our own century that employ (or employed) virtually all the technologies and economic strategies that have been developed since the beginning of the Upper Paleolithic. Given the general pattern of changelessness that characterizes the bulk of the Paleolithic archaeological record, then, it is hard not to conclude that there was but one truly great leap (forward?) in human evolution: the one that gave rise to our own species, *Homo sapiens*. If you'd been around at any earlier stage of human evolution, with some knowledge of the past, you might have been able to predict with reasonable accuracy what might be coming up next. *Homo sapiens*, however, is emphatically not an organism that does what its predecessors did, only a little better; it's something very—and potentially very dangerously—different. Something extraordinary, if totally fortuitous, happened with the birth of our species. And although the human biological past stretches back over five million poorly known years or more, it is the nature of that very recent yet still obscure happening that poses the true enigma of human evolution.

EPILOGUE

I have just said that in the case of the emergence of our species *Homo sapiens*, perhaps for the first time, past performance has been no guarantee—or, at least, no predictor—of future results. But can we hazard anything about our future as a species from looking at the general rules that govern the evolutionary process—for, extraordinary as we are, there's no reason to believe that we have become emancipated from them? Despite a limitless amount of historical evidence to the contrary, human beings appear to have enormous difficulty in ridding themselves of the notion of their own potential perfectibility. One way in which the perfectibility of mankind might be achieved is particularly beloved of science fiction writers, who regularly proffer visions of future people with huge brains (if generally frailer bodies), whose rational qualities actually manage successfully to assert themselves over their more atavistic, emotional, instincts.

But is it in fact at all reasonable to contemplate a future in which evolution will, as it were, ride in on a white horse to rescue us from ourselves? A moment's thought shows how unlikely this is. Everything that we know rather than assume about the evolutionary process indicates that population fragmentation (supplemented by genetic innovation and followed by speciation) is a prerequisite for the accumulation of any significant evolutionary change in mammals. Yet we are living at a time when the human population is becoming denser by leaps and bounds, and when individual mobility has become incomparably greater than ever before. *Homo sapiens* today is in a mode of intermixing rather than of differentiation, and the conditions for significant evolutionary change simply don't exist—and won't, short of some all-too-imaginable calamity. In the demonstrable absence of perfectibility, if that calamity is not to occur we shall have to learn to live with ourselves as we are. Fast.

BIBLIOGRAPHY

No bibliography of reasonable length could include all of the publications consulted in the preparation of this work. The following list nonetheless identifies most of the major books and articles referred to (with an emphasis on those most accessible to the general reader), including those from which direct quotations have been taken. Volumes with asterisks are easily accessible reviews and/or contain extensive bibliographies. For convenience, references from each chapter of this book are listed together.

Chapter 1

Buffon, G. L. 1779. Les Epoques de la Nature. [Edited by J. Roger, *Mém. Mus. Nat. Hist. Natl., Paris*, C, 10: 1–343, 1962.]

Chambers, R. 1844. *Vestiges of the Natural History of Creation*. London: Churchill.

Cuvier, G. 1812. *Récherches sur les Ossemens Fossiles de Quadrupèdes, où l'on Rétablit les Caractères de Plusieurs Espèces d'Animaux que les Révolutions du Globe Paroissent Avoir Détruites*. Paris: Deterville.

*Grayson, D. K. 1983. *The Establishment of Human Antiquity*. New York: Academic Press.

Grayson, D. K. 1990. The provision of time depth for paleoanthropology. *Geol. Soc. Amer. Spec. Paper* 242: 1–13.

Lamarck, J.-B. 1809. *Philosophie Zoologique*. [*The Zoological Philosophy*, London: Macmillan, 1914 (translation).]

Lyell, C. 1830/32. *Principles of Geology, Being an Attempt to Explain the Former Changes of the Earth's Surface by Reference to Causes Now in Operation*. Vol. 1, 1830; Vol. 2, 1832. London: John Murray.

Mayer, F. 1864. Über die Fossilen Uberreste eines Menschlichen Schädels und Skeletes in einer Felsenhöhle des Düssel—oder Neander—Thales. *Arch. Inst. Physiol.* for 1864, 1–26.

*Mayr, E. 1982. *The Growth of Biological Thought: Diversity, Evolution, and Inheritance*. Cambridge, MA: Belknap Press.

Schaaffhausen, H. 1853. Über Beständigkeit und Umwandlung der Arten. *Verh. Naturhist. Vereins Preussischen Rheinlande und Westphalens* 10: 420–451.

Schaaffhausen, H. 1858. Zur Kentniss der ältesten Rassenschädel. [On the crania of the most ancient races of man]. *Nat. Hist. Rev.* 1: 155–76, 1861 (translation, with introduction by G. Busk).]

Schmerling, P. C. 1833–34. *Recherches sur les Ossemens Fossiles Découvertes dans les Cavernes de la Province de Liège.* Vol. 1, 1833; Vol. 2, 1834. Liège: Collardin.

Chapter 2

*Bahn, P., and J. Vertut. 1988. *Images of the Ice Age.* New York: Facts on File.

*Bowler, P. J. 1986. *Theories of Human Evolution: A Century of Debate, 1844–1944.* Baltimore: Johns Hopkins University Press.

Busk, G. 1864. Pithecan priscoid man from Gibraltar. *The Reader,* 23 July 1864.

Darwin, C. R. 1859. *On the Origin of Species by Means of Natural Selection: Or the Preservation of Favoured Races in the Struggle for Life.* London: John Murray.

Darwin, C. R. 1871. *The Descent of Man in Relation to Sex.* London: John Murray.

de Mortillet, G. 1883. *Le Préhistorique: Antiquité de l'Homme.* Paris: Reinwald.

Haeckel, E. 1868. *Natürliche Schöpfungsgeschichte [The History of Creation.* New York: Appleton, 1892 (translation).]

Huxley, T. H. 1863. *Evidence as to Man's Place in Nature.* London: Williams & Norgate.

King, W. 1863. The Neanderthal skull. *Anthrop. Rev.,* 1: 393–94.

Lartet, E. 1861. Nouvelles recherches sur la coexistance de l'homme et des grands mammifères fossiles. [New researches concerning the co-existence of man with the great fossil mammals, regarded as characteristic of the latest geological period. *Nat. Hist. Review, for 1862,* 53–71. (translation).]

Lartet, E., and H. Christy. 1865. *Reliquiae Aquitaniae.* [*R.A.: Being Contributions to the Archaeology and Palaeontology of Périgord and the Adjoining Provinces of Southern France.* London: Williams & Norgate, 1875 (translation).]

Lubbock, J. 1865. *Prehistoric Times: As Illustrated by Ancient Remains and the Manners and Customs of Modern Savages.* London: Williams & Norgate.

*Mayr, E. 1982. *The Growth of Biological Thought: Diversity, Evolution, and Inheritance.* Cambridge, MA: Belknap Press.

*Milner, R. 1990. *The Encyclopedia of Evolution: Humanity's Search for Its Origins.* New York: Facts on File.

*Theunissen, B. 1988. *Eugène Dubois and the Ape-Man from Java: The History of the First "Missing Link" and Its Discoverer.* Dordrecht/Boston: Kluwer Academic.

Virchow, R. 1872. Untersuchung des Neanderthal-Schädels. *Zool.-Ethnol.* 4: 157–65.

Wallace, A. R. 1858. On the tendency of species to depart indefinitely from the original type. *Proc. Linn. Soc. London (Zoology)* 3: 53–62.

Chapter 3

Cunningham, D. J. 1895. The place of "Pithecanthropus" on the genealogical tree. *Nature* 53: 269.

Dubois, E. 1891. Palaeontologische onderzoekingen op Java. *Verslag van het Mijnwezen* 9: 12–15.

Dubois, E. 1894. *Pithecanthropus erectus,* eine menschenähnliche Uebergangsform aus Java. Batavia, Landesdrukkerei.

Fraipont, J., and M. Lohest. 1886. La race humaine de Néanderthal ou de Canstadt en Belgique. *Arch. Biol.* 7: 587–755.

*Reader, J. 1981. *Missing Links: The Hunt for Earliest Man.* Boston: Little, Brown.

Schwalbe, G. 1899. Studien über Pithecanthropus erectus Dubois. *Morphol. Anthropol.* 1: 16–228.

Schwalbe, G. 1900. Der Neanderthalschädel. *Jahrb. Verh. Altetsfr. Rheinlande* 106: 1–72.

*Theunissen, B. 1988. *Eugène Dubois and the Ape-Man from Java: The History of the First "Missing Link" and Its Discoverer.* Dordrecht/Boston: Kluwer Academic.

Chapter 4

Boule, M. 1911–13. L'homme fossile de La Chapelle-aux-Saints. *Annales de Paléontologie* 6 (1911): 1–64; 7 (1912): 65–208; 8 (1913): 209–279.

Dawson, C., and A. S. Woodward. 1913. On the discovery of a Palaeolithic human skull and mandible in a flint-bearing gravel overlying the Wealden (Hastings Beds) at Piltdown, Fletching, (Sussex) *Quart. Jour. Geol. Soc. London* 69: 117–51.

DeVries, H. 1901. Die Mutationstheorie. Versuche und Beobachtungen der Arten im Pflanzreich, Vol. 1. [*Species and Varieties: Their Origin by Mutation.* Chicago: Open Court Publishing, 1906). (translation).]

Hrdlička, A. 1927. The Neanderthal phase of man. *Jour. Roy. Anthropol. Inst.* 57: 249–274.

Hrdlička, A. 1930. *The Skeletal Remains of Early Man.* Smithsonian Misc. Coll. 83 (entire vol.).

Huxley, T. H. 1893/94. *Collected Essays.* Vol. 2, *Darwiniana*, 1893; Vol. 7, *Man's Place in Nature*, 1893. London: Macmillan.

Keith, A. 1913. The Piltdown skull and brain cast. *Nature* 92: 107–9, 197–99, 292, 345–46.

*Mayr, E. 1982. *The Growth of Biological Thought: Diversity, Evolution, and Inheritance.* Cambridge, MA: Belknap Press.

Mendel, G. 1866. Versuche über Pflanzen-hybriden. *Verh. Natur. Vereins Brnn* 4: 3–57.

*Reader, J. 1981. *Missing Links: The Hunt for Earliest Man.* Boston: Little, Brown.

Schoetensack, O. 1908. *Der Unterkiefer des* Homo heidelbergensis *aus den Sanden von Mauer bei Heidelberg.* Leipzig: Engelmann.

Smith, G. E. 1913. The Piltdown skull and brain cast. *Nature* 92: 267–268, 318–319.

*Spencer, F. 1990. *Piltdown: A Scientific Forgery.* London: Natural History Museum/ Oxford University Press.

Trémaux, P. 1865. *Origine et Transformations de l'Homme et des Autres Etres.* Paris: Hachette.

Chapter 5

Anon. 1938. Pithecanthropus erectus—"The Ape-Man of Java." *Carnegie Inst. News Serv. Bull.* 4(27): 227–32.

Black, D. 1927. On a lower molar hominid tooth from the Chou Kou Tien deposit. *Palaeont. Sinica*, ser. D, 7: 1–29.

Black, D. 1931. Evidences of the use of fire by *Sinanthropus*. *Bull. Geol. Soc. China* 11: 107–8.

Boule, M. 1929. Le *Sinanthropus*. *L'Anthropologie* 39: 455–60.

Boule, M. 1937. Le Sinanthrope. *L'Anthropologie* 47: 1–22.

*Clark, W. E. Le Gros. 1967. *Man-Apes or Ape-Men? The Story of Discoveries in Africa.* New York: Holt, Rinehart and Winston.

Dart, R. A. 1925. *Australopithecus africanus*: the man-ape of South Africa. Nature 115: 195–99.

Dubois, E. 1933. The shape and the size of the brain in Sinanthropus and in Pithecanthropus. *Proc. Kon. Ned. Akad. Wet.* 36: 415–23.

Koenigswald, G. H. R. von. 1938. Ein Neuer Pithecanthropus-Schädel. *Proc. Akad. Sci. Amst.* 41: 185–192.

Koenigswald, G. H. R. von, and F. Weidenreich. 1939. The relationship between Pithecanthropus and Sinanthropus. *Nature* 144: 926–27.

Oppenoorth, W. F. F. 1932. *Homo (Javanthropus) soloensis*, een plistoceene Mensch von Java. *Wet. Meded. Dienst. Mijnb. Ned.-Oest. Indië* 20: 49–75.

Osborn, H. F. 1916. *Men of the Old Stone Age.* New York: Charles Scribner's Sons.

*Reader, J. 1981. *Missing Links: The Hunt for Earliest Man.* Boston: Little, Brown.

*Shapiro, H. 1974. *Peking Man: The Discovery, Disappearance and Mystery of a Priceless Scientific Treasure.* New York: Simon and Schuster.

Weidenreich, F. 1938. *Pithecanthropus* and *Sinanthropus*. Nature 141: 378–79.

Weidenreich, F. 1939. Six lectures on Sinanthropus pekinensis and related problems. *Bull. Geol. Soc. China* 19: 1–110.

Weidenreich, F. 1943. The skull of *Sinanthropus pekinensis*: a comparative study on a primitive hominid skull. *Palaeont. Sinica*, new ser. D, 10: 1–291.

Weidenreich, F. 1947. Facts and speculations concerning the origin of Homo sapiens. *Amer. Anthropol.* 49: 187–203.

Wernert, P. 1948. *Le culte des crânes à l'époque Paléolithique.* Histoire Générale des Réligions, edited by M. Gorce and R. Mortier, Vol. 1. Paris: Quillet.

Woodward, A. S. 1921. A new cave man from Rhodesia, South Africa. *Nature* 108: 371–72.

Zdansky, O. 1927. Preliminary notice on two teeth of a hominid from a cave in Chihli (China). *Bull. Geol. Soc. China* 5: 281–284.

Chapter 6

Ardrey, R. 1961. *African Genesis.* New York: Atheneum.

Broom, R. 1936. A new fossil anthropoid skull from South Africa. *Nature* 138: 486–88.

Broom, R. 1937. The Sterkfontein ape. *Nature* 139: 326.

Broom, R. 1949. Another new type of fossil ape-man. *Nature* 163: 57.

Broom, R., and J. T. Robinson. 1949. Man contemporaneous with Swartkrans apeman. *Amer. Jour. Phys. Anthropol.* 8: 151–56.

Broom, R., and J. T. Robinson. 1950. Further evidence of the structure of the Sterkfontein Ape-Man *Plesianthropus*. *Transvaal Mus. Memoir* 4(1): 8–83.

Broom, R., and G. W. H. Schepers. 1946. The South African fossil ape-men. The Australopithecinae. *Transvaal Mus. Memoir* 2: 1–271.

*Clark, W. E. Le Gros. 1967. *Man-Apes or Ape-Men? The Story of Discoveries in Africa.* New York: Holt, Rinehart and Winston.

Dart, R. A. 1948. The Makapansgat proto-human *Australopithecus prometheus*. *Amer. Jour. Phys. Anthropol.* 6: 259–84.

Dart, R. A. 1957. The osteodontokeratic culture of *Australopithecus prometheus*. *Transvaal Mus. Memoir* 10: 1–105.

Gregory, W. K. 1939. The South African fossil man-apes and the origin of the human dentition. *Jour. Amer. Dental Assoc.* 26: 645

Hooton, E. A. 1931. *Up from the Ape.* New York, Macmillan. 2nd ed., 1946.

Keith, A. 1930. *New Discoveries Relating to the Antiquity of Man.* London: Williams & Norgate.

Keith, A. 1948. *A New Theory of Human Evolution.* London: Watts.

Koenigswald, G. H. R. von. 1942. *The South African Man-Apes and Pithecanthropus.* Carnegie Inst. Publ. 530: 205–22.

Koenigswald, G. H. R. von. 1956. *Meeting Prehistoric Man.* New York: Harper & Bros.

McCown, T., and A. Keith. 1939. *The Stone Age of Mount Carmel*, Vol. 2. Oxford: Clarendon Press.

*Reader, J. 1981. *Missing Links: The Hunt for Earliest Man.* Boston: Little, Brown.

Schepers, G. W. H. 1950. The brain casts of the recently discovered *Plesianthropus* skulls. *Transvaal Mus. Memoir* 4(2): 85–117.

*Stringer, C. B., and C. Gamble. 1993. *In Search of the Neanderthals: Solving the Puzzle of Human Origins.* London: Thames and Hudson.

*Trinkaus, E., and P. Shipman. 1993. *The Neanderthals: Changing the Image of Mankind.* New York: Knopf.

Weidenreich, F. 1948. About the morphological character of the Australopithecinae skull. In *Rob't Broom, Memorial Volume*, Spec. Publ. Roy. Soc. S. Afr.: 153–58.

Weinert, H. 1950. Über die Neuen Vor- und Frühmenschenfunde aus Afrika, Java, China und Frankreich. *Zeitschr. Morph. Anthrop.* 42: 113–48.

*Willis, D. 1992. *The Leakey Family: Leaders in the Search for Human Origins.* New York: Facts on File.

Chapter 7

Arambourg, C. 1955. A recent discovery in human paleontology: *Atlanthropus* of Ternifine (Algeria). *Amer. Jour. Phys. Anthropol.* 13: 191–202.

Arambourg, C. 1955. Sur l'attitude, en station verticale, de Néanderthaliens. *C. R. Acad. Sci. Paris* 240: 804–6.

Dobzhansky, T. 1937. *Genetics and the Origin of Species.* New York: Columbia University Press.

Dobzhansky, T. 1944. On species and races of living and fossil man. *Amer. Jour. Phys. Anthropol.* 2: 251–65.

*Eldredge, N. 1985. *Time Frames: The Rethinking of Darwinian Evolution and the Theory of Punctuated Equilibria.* New York: Simon and Schuster.

Howell, F. C. 1951. The place of Neanderthal Man in human evolution. *Amer. Jour. Phys. Anthropol.* 9: 379–416.

Howell, F. C. 1952. Pleistocene glacial ecology and the evolution of "Classic Neanderthal" man. *Southwest. Jour. Anthrop.* 8: 377–410.

Mayr, E. 1950. Taxonomic categories in fossil hominids. *Cold Spring Harbor Symp. Quant. Biol.* 15: 109–18.

*Mayr, E. 1982. *The Growth of Biological Thought: Diversity, Evolution, and Inheritance.* Cambridge, MA: Belknap Press.

Movius, H. 1973. The Abri Pataud program of the French Upper Paleolithic in retrospect. In *Archaeological Research in Retrospect,* G. Willey, ed. Cambridge, MA: Winthrop, 87–116.

Simpson, G. G. 1944. *Tempo and Mode in Evolution.* New York: Columbia University Press.

Solecki, R. S. 1971. *Shanidar: The First Flower People.* New York: Knopf.

Straus, W. L., and J. E. Cave. 1957. Pathology and the posture of Neanderthal Man. *Quart. Rev. Biol.* 32: 348–63.

*Stringer, C. B., and C. Gamble. 1993. *In Search of the Neanderthals: Solving the Puzzle of Human Origins.* London: Thames and Hudson.

*Trinkaus, E., and P. Shipman. 1993. *The Neanderthals: Changing the Image of Mankind.* New York: Knopf.

Wright, S. 1932. The roles of mutation, inbreeding, crossbreeding, and selection in evolution. *Proc. 6th Intl. Congr. Genetics* 1: 356–66.

Chapter 8

Brace, C. L. 1964. The fate of the "Classic" Neanderthals: a consideration of hominid catastrophism. *Curr. Anthropol.* 5: 3–43.

Day, M. H. and J. R. Napier. 1964. Hominid fossils from Bed I, Olduvai Gorge, Tanganyika. Fossil foot bones. *Nature* 201: 967–70.

Koenigswald, G. H. R. von, W. Gentner, and H. J. Lippolt. 1961. Age of the basalt flow at Olduvai, East Africa. *Nature* 192: 720–21.

Leakey, L. S. B. 1959. A new fossil skull from Olduvai. *Nature* 184: 491–93.

Leakey, L. S. B. 1961. New finds at Olduvai Gorge. *Nature* 189: 649–50.

Leakey, L. S. B., J. F. Evernden, and G. H. Curtis. 1961. Age of Bed I, Olduvai Gorge, Tanganyika. *Nature* 191: 478–479.

Leakey, L. S. B., and M. D. Leakey. 1964. Recent discoveries of fossil hominids in Tanganyika: at Olduvai and near Lake Natron. *Nature* 202: 5–7.

Leakey, M. D. 1966. A review of the Oldowan culture from Olduvai Gorge, Tanzania. *Nature* 210: 462–66.

*Leakey, M. D. 1971. *Olduvai Gorge,* Vol. 3. Cambridge: Cambridge University Press.

Oakley, K. P. 1949. *Man the Toolmaker*. London: British Museum (Natural History). (Several later editions.)

Pilbeam, D. R., and E. L. Simons. 1965. Some problems of hominid classification. *Amer. Scientist* 53: 237–59.

*Reader, J. 1981. *Missing Links: The Hunt for Earliest Man*. Boston: Little, Brown.

Robinson, J. T. 1960. The affinities of the new Olduvai australopithecine. *Nature* 186: 456–58.

Robinson, J. T. 1966. The distinctiveness of *Homo habilis*. *Nature* 209: 957–60.

Tobias, P. V. 1963. Cranial capacity of Zinjanthropus and other australopithecines. *Nature* 197: 743–46.

*Tobias, P. V. 1967. *Olduvai Gorge*, Vol. 2. Cambridge: Cambridge University Press.

*Tobias, P. V. 1991. *Olduvai Gorge*, Vol. 4. Cambridge: Cambridge University Press.

Tobias, P. V., and G. H. R. von Koenigswald. 1964. Comparison between the Olduvai hominines and those of Java and some implications for hominid phylogeny. *Nature* 204: 515–18.

*Willis, D. 1992. *The Leakey Family: Leaders in the Search for Human Origins*. New York: Facts on File.

Chapter 9

Andrews, P., and J. E. Cronin. 1982. The relationships of *Sivapithecus* and *Ramapithecus* and the evolution of the orang-utan. *Nature* 297: 541–46.

Andrews, P., and I. Tekkaya. 1980. A revision of the Turkish Miocene hominoid *Sivapithecus meteai*. *Palaeontology* 23: 83–95.

*Conroy, G. C. 1990. *Primate Evolution*. New York: Norton.

Goodman, M. Immunochemistry of the primates and primate evolution. *Ann. N. Y. Acad. Sci.* 102: 219–34.

Hrdlička, A. 1935. The Yale fossils of anthropoid apes. *Amer. Jour. Sci.* 229: 533–38.

Leakey, L. S. B. 1962. A new lower Pliocene fossil primate from Kenya. *Ann. Mag. Nat. Hist.*, Ser 13, 4: 689–96.

Leakey, L. S. B. 1967. An early Miocene member of Hominidae. *Nature* 213: 155–63.

*Lewin, R. 1987. *Bones of Contention: Controversies in the Search for Human Origins*. New York: Simon and Schuster.

Lewis, G. E. 1934. Preliminary notice of man-like apes from India. *Amer. Jour. Sci.* 27: 161–79.

Nuttall, G. H. F. 1904. *Blood Immunity and Blood Relationship*. Cambridge: Cambridge University Press.

Pilbeam, D. R. 1979. Recent finds and interpretations of Miocene hominoids. *Ann. Rev. Anthropol.* 8: 333–52.

Sarich, V. M., and A. C. Wilson. 1966. Quantitative immunochemistry and the evolution of primate albumins: micro-complement fixation. *Science* 154: 1563–66.

Sarich, V. M. and A. C. Wilson. 1967. Immunological time scale for hominid evolution. *Science* 158: 1200–1203.

Sarich, V. M. and A. C. Wilson. 1967. Rates of albumin evolution in primates. *Proc. Nat. Acad. Sci.* 58: 142–48.

*Schwartz, J. H. 1987. *The Red Ape: Orang-utans and Human Origins*. Boston: Houghton Mifflin.

Simons, E. L. 1961. The phyletic position of *Ramapithecus*. *Peabody Mus. Postilla* 57: 1–9.

Simons, E. L. 1964. On the mandible of *Ramapithecus*. *Proc. Nat. Acad. Sci.* 51: 528–35.

Simons, E. L., and D. R. Pilbeam. 1965. Preliminary revision of the Dryopithecinae (Pongidae, Anthropoidea). *Folia Primatol.* 3: 81–152.

Vogel, C. 1975. Remarks on the reconstruction of the dental arcade of *Ramapithecus*. In *Paleoanthropology. Morphology and Paleoecology*, R. H. Tuttle, ed. Chicago: Aldine, 87–98.

Walker, A. C., and P. Andrews. 1973. Reconstruction of the dental arcade of *Ramapithecus wickeri*. *Nature* 244: 313–314.

Chapter 10

Alexeev, V. P. 1986. *The Origin of the Human Race*. Moscow: Progress Publishers.

*Binford, L. R. 1981. *Bones: Ancient Men and Modern Myths*. Orlando, FL: Academic Press.

Brace, C. L. 1964. The fate of the "Classic" Neanderthals: a consideration of hominid catastrophism. *Curr. Anthropol.* 5: 3–43.

Brace, C. L. and A. Montagu. 1965. *Human Evolution: An Introduction to Biological Anthropology*. New York: Macmillan.

Brown, F. H., and C. S. Feibel. 1986. Revision of lithostratigraphic nomenclature of the Koobi Fora region, Kenya. *Jour. Geol. Soc.* 143: 297–310.

Day, M. H., R. E. F. Leakey, A. C. Walker, and B. A. Wood. 1976. New hominids from East Turkana, Kenya. *Amer. Jour. Phys. Anthropol.* 45: 369–436.

Feibel, C. S., F. H. Brown, and I. MacDougall. 1989. Stratigraphic context of fossil hominids from the Omo group deposits: Northern Turkana Basin, Kenya and Ethiopia. *Amer. Jour Phys. Anthropol.* 78: 595–622.

Groves, C. P., and V. Mazak. 1975. An approach to the taxonomy of the Hominidae: gracile Villafranchian hominids of Africa. *Casopis pro Mineralogii Geologii* 20: 225–47.

Howell, F. C. 1969. Remains of Hominidae from Pliocene/Pleistocene formations in the lower Omo basin, Ethiopia. *Nature* 223: 1234–39.

*Howell, F. C. 1978. Hominidae. In *Evolution of African Mammals*, V. J. Maglio and H. B. S. Cooke, eds. Cambridge, MA: Harvard University Press, 154–248.

Isaac, G. L. 1978. The food-sharing behavior of proto-human hominids. *Scientific American* 238: 90–108.

Isaac, G. L. 1983. Bones in contention: competing explanations for the juxtaposition of Early Pleistocene artefacts and faunal remains. In *Animals and Archaeology: Hunters and Their Prey*, J. Clutton-Brock and G. Grigson, eds. Oxford: British Archaeological Reports, 3–19.

*Johanson, D. J. and M. A. Edey. 1981. *Lucy: The Beginnings of Humankind*. New York: Simon and Schuster.

Leakey, R. E. F. 1970. New hominid remains and early artefacts from Northern Kenya. *Nature* 226: 226–28.

Leakey, R. E. F. 1971. Further evidence of Lower Pleistocene hominids from East Rudolf, Kenya. *Nature* 231: 241–45.

Leakey, R. E. F. 1972. Further evidence of Lower Pleistocene hominids from East Rudolf, North Kenya. *Nature* 237: 264–69.

Leakey, R. E. F. 1973. Further evidence of Lower Pleistocene hominids from East Rudolf, North Kenya 1972. *Nature* 242: 170–73.

Leakey, R. E. F. 1974. Further evidence of Lower Pleistocene hominids from East Rudolf, North Kenya 1973. *Nature* 248: 653–56.

Leakey, R. E. F., and A. C. Walker. 1976. *Australopithecus, Homo erectus* and the single species hypothesis. *Nature* 261: 572–74.

*Lewin, R. 1987. *Bones of Contention: Controversies in the Search for Human Origins.* New York: Simon and Schuster.

*Willis, D. 1992. *The Leakey Family: Leaders in the Search for Human Origins.* New York: Facts on File.

Wolpoff, M. H. 1980. *Paleoanthropology.* New York: Knopf.

*Wood, B. 1991. *Koobi Fora Research Project*, Vol. 4. Oxford: Clarendon Press.

Chapter 11

Clark J. D., B. Asfaw, G. Assefa, J. W. K. Harris, H. Kurashina, R. C. Walter, T. D. White, and M. A. J. Williams. 1984. Palaeoanthropological discoveries in the Middle Awash Valley, Ethiopia. *Nature* 307: 423–28.

Conroy, G. C., C. J. Jolly, D. Cramer, and J. E. Kalb. 1978. Newly discovered fossil hominid skull from the Afar Depression, Ethiopia. *Nature* 275: 67–70.

Day, M. H., M. D. Leakey, and C. Magori. 1980. A new hominid skull (LH 18) from the Ngaloba Beds, Laetoli, Northern Tanzania. *Nature* 284: 55–56.

Falk, D. 1993. A good brain is hard to cool. *Natural History* 102(8): 65.

*Johanson, D. C., and M. Edey. 1981. *Lucy: The Beginnings of Humankind.* New York: Simon and Schuster.

Johanson, D. C., and M. Taieb. 1976. Plio-Peistocene hominid discoveries in Hadar, Ethiopia. *Nature* 260: 293–97.

Johanson, D. C., and T. D. White. 1979. A systematic assessment of early African hominids. *Science* 202: 321–30.

Johanson, D. C., T. D. White, and Y. Coppens. 1978. A new species of the genus *Australopithecus* (Primates: Hominidae) from the Pliocene of eastern Africa. *Kirtlandia* 28: 1–14.

Johanson, D. C., and others. 1982. Pliocene hominid fossils from Hadar, Ethiopia. *Amer. Jour. Phys. Anthropol.* 57(4): 373–724.

*Leakey, M. D. and J. M. Harris, eds. 1987. *Laetoli: A Pliocene Site in Northern Tanzania.* Oxford: Clarendon Press.

Olson, T. 1981. Basicranial morphology of the extant hominoids and Pliocene hominids: the new material from the Hadar Formation, Ethiopia, and its significance in early human evolution and taxonomy. In *Aspects of Human Evolution*, C. B. Stringer, ed. London: Taylor and Francis, 99–128.

Taieb, M., D. C. Johanson, Y. Coppens, and J. L. Aronson. 1976. Geological and palaeontological background of Hadar hominid site, Afar, Ethiopia. *Nature* 260: 289–93.

Tobias, P. V. 1980. *"Australopithecus afarensis"* and *A. africanus*: critique and alternative hypothesis. *Palaeont. Africana* 23: 1–17.

Weinert, H. 1950. Über die Neuen Vor- und Frühmenschenfunde aus Afrika, Java, China und Frankreich. *Zeitschr. Morph. Anthrop.* 42: 113–48.

Wheeler, P. 1993. Human ancestors walked tall, stayed cool. *Natural History* 102: 65–67.

White, T. D. 1977. New fossil hominids from Laetoli, Tanzania. *Amer. Jour. Phys. Anthrop.* 53: 197–230.

White, T. D. 1980. Additional hominid specimens from Laetoli, Tanzania. *Amer. Jour. Phys.* Anthropol. 53: 487–504.

Chapter 12

Eldredge, N. 1971. The allopatric model and phylogeny in Paleozoic invertebrates. *Evolution* 25: 156–67.

*Eldredge, N. 1985. *Time Frames: The Rethinking of Darwinian Evolution and the Theory of Punctuated Equilibria*. New York: Simon and Schuster.

*Eldredge, N., and J. Cracraft. 1980. *Phylogenetic Patterns and the Evolutionary Process*. New York: Columbia University Press.

Eldredge, N., and S. J. Gould. 1972. Punctuated equilibria: an alternative to phyletic gradualism. In *Models in Paleobiology*, T. J. M. Schopf, ed. San Francisco: Freeman Cooper, pp. 82–115.

Eldredge, N., and I. Tattersall. 1975. Evolutionary models, phylogenetic reconstruction, and another look at hominid phylogeny. In *Approaches to Primate Paleobiology*, F. S. Szalay, ed. Basel: Karger, 218–42.

Hennig, W. 1950. Grundzüge einer Theorie der phylogenetischen Systematik. [*Phylogenetic Systematics*. Urbana: University of Illinois Press, 1966 (translation).]

Mayr, E. 1942. *Systematics and the Origin of Species*. New York: Columbia University Press. Reprint, 1982.

*Otte, D. and J. A. Endler, eds. 1989. *Speciation and Its Consequences*. Sunderland, MA: Sinauer Associates.

Tattersall, I., and N. Eldredge. 1977. Fact, theory and fantasy in human paleontology. *Amer. Scientist* 65: 204–11.

Chapter 13

Antunes, M. T., and A. Santinho Cunha. 1991. Neanderthalian remains from Figueira Brava Cave, Portugal. *Géobios* 25: 681–92.

Arsuaga, J.-L., I. Martinez, A. Gracia, J.-M. Carretero and E. Carbonell. 1993. Three new human skulls from the Sima de los Huesos Middle Pleistocene site in Sierra de Atapuerca, Spain. *Nature* 362: 534–37.

Clarke, R. J. 1990. The Ndutu cranium and the origin of *Homo sapiens*. *J. Hum. Evol.* 19: 699–736.

*Day, M. H. 1988. *Guide to Fossil Man*, 4th ed. Chicago: University of Chicago Press.

Deacon, H. 1992. Southern Africa and modern human origins. *Phil. Trans. Roy. Soc. Lond.* B1280: 177–84.

*Gowlett, J. 1984. *Ascent to Civilization: The Archaeology of Early Man*. New York: Knopf.

*Klein, R. G. 1989. *The Human Career: Human Biological and Cultural Origins*. Chicago: University of Chicago Press.

Rightmire, G. P. 1990. *The Evolution of* Homo erectus: *Comparative Anatomical Studies of an Extinct Human Species*. New York: Cambridge University Press.

Singer, R., and J. Wymer. 1982. *The Middle Stone Age at Klasies River Mouth in South Africa*. Chicago: University of Chicago Press.

*Stringer, C. B., and C. Gamble. 1993. *In Search of the Neanderthals: Solving the Puzzle of Human Origins*. London, Thames and Hudson.

Stringer, C. B., F. C. Howell, and J. K. Melentis. 1979. The significance of the fossil hominid skull from Petralona, Greece. *J. Arch. Sci.* 6: 235–53.

*Tattersall, I. 1993. *The Human Odyssey: Four Million Years of Human Evolution*. New York: Prentice Hall.

*Tattersall, I., E. Delson, and J. A. Van Couvering, eds. 1988. *Encyclopedia of Human Evolution and Prehistory*. New York: Garland.

Chapter 14

Andrews, P. 1984. An alternative interpretation of the characters used to define *Homo erectus*. *Cour. Forsch. Inst. Senckenberg*, 69: 167–75.

Brown, F., J. Harris, R. Leakey, and A. Walker. 1985. Early *Homo erectus* skeleton from west Lake Turkana, Kenya. *Nature*, 316: 788–92.

*Grine, F. E., ed. 1988. *Evolutionary History of the "Robust" Australopithecines*. New York: Aldine de Gruyter.

Hughes, A. R., and P. V. Tobias. 1977. A fossil skull probably of the genus *Homo* from Sterkfontein, Transvaal. *Nature*, 265: 310–12.

Johanson, D. C., F. T. Masao, G. G. Eck, T. D. White, R. C. Walter, W. H. Kimbel, B. Asfaw, P. Manega, P. Ndessokia, and G. Suwa. 1987. New partial skeleton of *Homo habilis* from Olduvai Gorge, Tanzania. *Nature*, 327: 205–9.

*Johanson, D. C. and J. Shreeve. 1989. *Lucy's Child: The Discovery of a Human Ancestor*. New York: William Morrow.

*Leakey, R., and R. Lewin. 1992. *Origins Reconsidered: In Search of What Makes Us Human*. New York: Doubleday.

Stringer, C. B. 1986. The credibility of *Homo habilis*. In *Major Topics in Primate and Human Evolution*, B. Wood, L. Martin, and P. Andrews, eds. Cambridge: The University Press, 266–94.

Vrba, E. S. 1988. Late Pleistocene climatic events and hominid evolution. In *Evolutionary History of the "Robust" Australopithecines*, F. E. Grine, ed. New York: Aldine de Gruyter, 405–26.

Wood, B. 1985. *Australopithecus* and early *Homo* in East Africa. In *Ancestors: The Hard Evidence* E. Delson, ed. New York: Alan R. Liss, 206–14.

*Wood, B. 1991. Koobi Fora Research Project, Vol. 4: *Hominid Cranial Remains*. Oxford: Clarendon Press.

Wood, B. 1992. Origin and evolution of the genus *Homo*. *Nature*, 355: 783–90.

Chapter 15

*Binford, L. R. 1981. *Bones: Ancient Men and Modern Myths*. New York: Acadmic Press.

Binford, L. R. 1985. Human ancestors: changing views of their behavior. *J. Anthrop. Arch.* 4: 292–347.

Brain, C. K. 1970. New finds at the Swartkrans australopithecine site. *Nature*, 225: 1112–19.

Brain, C. K. 1976. A reinterpretation of the Swartkrans site and its remains. *S. Afr. J. Sci.* 72: 141–46.

*Brain, C. K. 1981. *The Hunters or the Hunted? An Introduction to African Cave Taphonomy*. Chicago: University of Chicago Press.

Brain, C. K., and A. Sillen. 1988. Evidence from the Swartkrans cave for the earliest use of fire. *Nature* 336: 464–66.

Bunn, H. T., and E. Kroll. 1986. Systematic butchery by Plio-Pleistocene hominids at Olduvai Gorge, Tanzania. *Curr. Anthrop.* 5: 431–52.

Clarke, R. J. 1988. A new *Australopithecus* cranium from Sterkfontein and its bearing on the ancestry of *Paranthropus*. In *Evolutionary History of the "Robust" Australopithecines*, F. E. Grine, ed. New York: Aldine de Gruyter, 285–92.

Clarke, R. J. 1990. Observations on some restored hominid specimens in the Transvaal Museum, Pretoria. In *From Apes To Angels: Essays in Honor of Phillip V. Tobias*, G. Sperber, ed. New York: Wiley-Liss, pp. 135–52.

Delson, E. 1988. Chronology of South African australopith site units. In *Evolutionary History of the "Robust" Australopithecines*, F. E. Grine, ed. New York: Aldine de Gruyter, 317–24.

Hughes, A. R., and P. V. Tobias. 1977. A fossil skull probably of *Homo* from Sterkfontein, Transvaal. *Nature* 265: 310–12.

Jones, P. 1981. Experimental implement manufacture and use: a case study from Olduvai Gorge. *Phil. Trans. Roy. Soc. Lond.* B292: 189–95.

Kay, R. F., and F. E. Grine. 1988. Tooth morphology, wear and diet in *Australopithecus* and *Paranthropus* from South Africa. In *Evolutionary History of the "Robust" Australopithecines*, F. E. Grine, ed. New York: Aldine de Gruyter, 427–48.

Laitman, J. T. 1988. Speech (Origins of). In *Encyclopedia of Human Evolution and Prehistory*, I. Tattersall, E. Delson, and J. A. Van Couvering, eds. New York: Garland, 539–40.

Leakey, M. D. 1970. Stone artefacts from Swartkrans. *Nature* 225: 1222–25.

*Schick, K. D., and N. Toth. 1993. *Making Silent Stones Speak*. New York: Simon and Schuster.

Susman, R. 1991. Who made the Oldowan tools? Fossil evidence for behavior in Plio-Pleistocene hominids. *J. Anthrop. Res.* 47: 129–52.

Tobias, P. V., and A. R. Hughes. 1969. The New Witwatersrand University excavation at Sterkfontein. *S. Afr. Arch. Bull.* 24: 158–69.

Toth, N., K. D. Schick, E. S. Savage-Rumbaugh, R. Sevcik, and D. Rumbaugh. 1993 Investigations in the stone tool-making and tool-using capabilities of a bonobo. *J. Arch. Sci.*, 20: 81–91.

Vrba, E. S. 1982. Biostratigraphy and chronology, based particularly on Bovidae, of southern hominid-associated assemblages: Makapansgat, Sterkfontein, Taung, Kromdraai, Swartkrans. In *Prétirage du Premier Congres. International de la Paléoontologie. Humaine*, Vol. 2, H. de Lumley and M.-A. de Lumley, eds. Nice: Centre. Nationale pour la Recherche Scientifique., 707–52.

Chapter 16

Bräuer, G. 1984. A craniological approach to the origin of anatomically modern *Homo sapiens* in Africa and implications for the appearance of modern Europeans. In *The Origins of Modern Humans: A World Survey of the Fossil Evidence*, F. H. Smith and F. Spencer, eds. New York: Alan R. Liss, 327–410.

Cann, R. L., M. Stoneking, and A. C. Wilson. 1987. Mitochondrial DNA and human evolution. *Nature* 325: 31–36.

Coon, C. S. 1962. *The Origin of Races*. New York: Knopf.

Howells, W. W. 1967. *Mankind in the Making*, rev. ed. New York: Doubleday.

Leakey, R. E. F., and R. Lewin. 1992. *Origins Reconsidered*. New York: Doubleday.

Nei, M., and A. K. Roychoudhury. 1974. Genic variation within and between the three major races of man, Caucasoids, Negroids and Mongoloids. *Amer. Jour. Hum. Genet.* 26: 421–43.

Stoneking, M., S. T. Sherry, A. J. Redd, and L. Vigilant. 1993. New approaches to dating suggest a recent age for the human mtDNA ancestor. *Phil. Trans. Roy. Soc. Lond.* B337: 167–75.

Stringer, C. B., and P. Andrews. 1988. Genetic and fossil evidence for the origin of modern humans. *Science*, 239: 1263–68.

*Stringer, C. B, and C. Gamble. 1993. *In Search of the Neanderthals: Solving the Puzzle of Human Origins*. London: Thames and Hudson.

Tattersall, I. 1986. Species recognition in human paleontology. *Jour. Hum. Evol.* 15: 165–75.

Tattersall, I. 1992. Species concepts and species identification in human evolution. *Jour. Hum. Evol.* 22: 341–49.

*Tattersall, I. 1993. *The Human Odyssey: Four Million Years of Human Evolution*. New York: Prentice Hall.

*Tattersall, I., E. Delson, and J. A. Van Couvering, eds. 1988. *Encyclopedia of Human Evolution and Prehistory*. New York: Garland.

Thorne, A. G., and M. H. Wolpoff. 1981. Regional continuity in Australasian Pleistocene hominid evolution. *Amer. Jour. Phys. Anthrop.* 55: 337–50.

Weidenreich, F. 1939. Six lectures on Sinanthropus pekinensis and related problems. *Bull. Geol. Soc. China* 19: 1–110.

Weidenreich, F. 1947. Facts and speculations concerning the origin of Homo sapiens. *Amer. Anthropol.* 49: 187–203.

Wolpoff, M. H., X. Wu, and A. G. Thorne. 1984. Modern *Homo sapiens* origins: a general theory of hominid evolution involving evidence from East Asia. In *The Origins of Modern Humans: A World Survey of the Fossil Evidence*, F. H. Smith and F. Spencer, eds. New York: Alan R. Liss, 411–83.

INDEX

Garrod, Dorothy, 83, 86
Garusi. *See* Laetoli site
Gaylenreuth cave, 8
Geikie, James, 47
Gene pool, 91
General lifeway, persistence of, 241
Genes, 41, 42
Genesis, biblical creation, 4, 5, 23
Genetics, 41, 42, 90, 91
Genotypes, 91
Gentner, Wolfgang, 112
Genus, 192, 193
Gibbon hypothesis, 40
Gibbons, 33, 36, 38, 40
Gibraltar, 9, 21, 22
Gigantopithecus, 121, 122, 214
Glacial phases, 47
Goodman, Morris, 123–125
Gorilla, 33, 36, 55, 74, 95,151, 218
Gould, Stephen Jay, 161, 164
Gracile *Australopithecus. See*
    Australopithecus africanus
Grades, 115
Granger, Walter, 59
Gravettian industry, 26i, 99
Greenfield, Leonard, 122
Gregory, William K., 71, 72
Grine, Fred, 195, 196, 202, 204
Groves, Colin, 139, 194

Hadar, environment, 155
    fossils, 142, 144, 145, 150–155, 158
    sites, 78i, 141, 142, 145, 146, 199
    Site 333, 145
Haeckel, Ernst, 24, 28, 29, 32, 35
Haile Selassie, Emperor, 128, 145
Half-life, definition of, 98
Half-life of ^{40}K, 111
Handaxe, 10i, 110, 177, 178, 188, 209,
    210, 240–244. *See also* Stone tools
Handaxes, from flakes, 243
"Handy Man," 113
Hardy-Weinberg principle, 42
Hartwig-Scherer, Sigrid, 191
Hat rack theory, 214

Hearths, 176, 182, 209
    appearance of, 244
Heidelberg. *See* Mauer site
Hellman, Milo, 71, 72
Hennig, Willi, 166
Hexian site, 173
Hierarchy, 4
Higher taxa, definition of, 91
Hill, Andrew, 149, 197
Home base, 136
Homeostasis, genetic, 42
Hominid catastrophism, 127
    definition, 126
Hominidae, 36, 71, 95, 96, 119, 120, 123,
    126, 152, 193, 230
Hominina, 126
Homininae, 71, 126
Hominini, 126
Hominization, 115
Hominoidea, 126
    definition, 35
Hominoids, 125
Homo, 27, 36, 46, 66, 96, 106, 113, 114,
    117, 122, 126, 127, 131–133, 144,
    151–154, 158,191, 193, 197, 202,
    205, 229, 230, 242
    definition of, 4
    early, 194
    morphological definition of, 193
Homo erectus, 35i, 76, 96, 109, 115–117,
    130, 133, 135, 138, 153, 168, 171,
    173–175, 187, 190, 191, 194, 201,
    202, 215–217, 219–221, 230, 232i,
    234i, 241, 242, 244
    Asia, new dates, 243n
Homo erectus, early African. See *Homo*
    ergaster
Homo ergaster, 139, 194, 202, 203, 207,
    209, 211, 231–234i, 238–241, 245
Homo habilis, 108i, 113–117, 130, 132–
    135, 153, 190–194, 202, 203, 208,
    230 , 232i, 232, 234i, 237–240
Homo heidelbergensis, 47, 219, 231– 234i,
    243, 244. *See also* Mauer jaw
Homo modjokertensis, 64

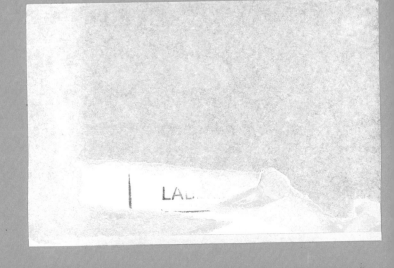

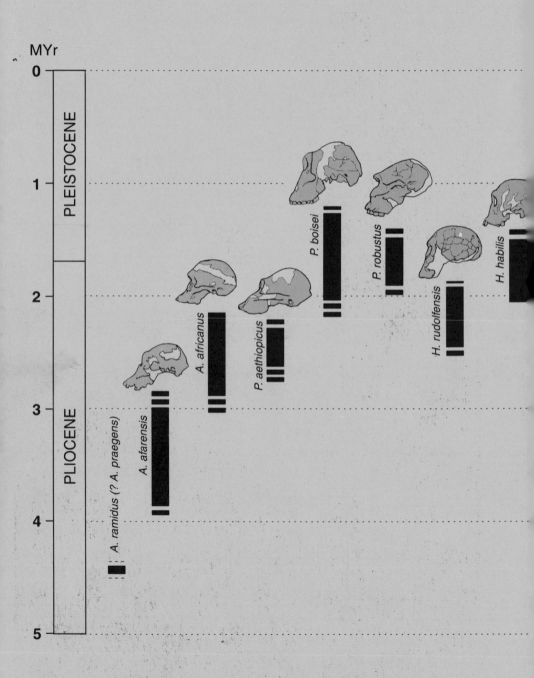

MYr

PLEISTOCENE

PLIOCENE

0

1

2

3

4

5

A. ramidus (? A. praegens)

A. afarensis

A. africanus

P. aethiopicus

P. boisei

P. robustus

H. rudolfensis

H. habilis